brilliant

meetings

meetings

What to know, do and say to have fewer, better meetings

Duncan Peberdy and Jane Hammersley

Harlow, England • London • New York • Boston • San Francisco • Toronto • Sydney • Singapore • Hong Kong
Tokyo • Seoul • Taipei • New Delhi • Cape Town • Madrid • Mexico City • Amsterdam • Munich • Paris • Milan

PEARSON EDUCATION LIMITED

Edinburgh Gate
Harlow CM20 2JE
Tel: +44 (0)1279 623623
Fax: +44 (0)1279 431059
Website: www.pearsoned.co.uk

First published in Great Britain in 2009

ISBN: 978–0–273–72182–6

British Library Cataloguing-in-Publication Data
A catalogue record for this book is available from the British Library

Library of Congress Cataloging-in-Publication Data
A catalog record for this book is available from the Library of Congress

10 9 8 7 6 5 4 3 2 1
13 12 11 10 09 Henry Ling Ltd., at the Dorset Press Dorchester, Dorset

Typeset in 10/14pt Plantin by 3
Printed and bound by Henry Ling Ltd, at the Dorset Press, Dorchester, Dorset

The publisher's policy is to use paper manufactured from sustainable forests.

Agenda

About the authors

Duncan Peberdy

Whilst I'm still not sure 'what I want to be when I grow up', an earlier career selling IT opened up an opportunity to write articles for several leading consumer technology publications. This was followed by my first novel – *Out Of Control.*

More recently I have worked with a leading UK University developing learning and teaching strategies using Multi-Display Environments. Such collaborative solutions have been successfully implemented now by leading global corporations.

I live in Droitwich Spa with my wife Elaine and our two grown-up children: Sophie and Tom. Photography and writing are my creative passions, and in old age I aspire to be 'Victor Meldrew'.

Jane Hammersley

My career path has been diverse and exciting, spanning both hardware and software environments, manufacturers and distributors all with a single common thread – Business Meetings.

A salesman by trade and author by a twist of fate, the inspiration for this book has derived from meeting experiences during the last 24 years, both in the workplace and as part of volunteer committees. Today, I am actively engaged with organisations – putting our 'Brilliant Meetings' strategies in to practice.

Leading a 'grown-up' life can be difficult, but my six-year-old daughter Eve offers me solace and inspiration, when needed, whilst my husband Mark challenges and loves me, the formula for our happy and fulfilled family life.

Together we have a real passion to make a difference to everyone who attends, arranges or leads meetings, to enable personal satisfaction to be achieved as well as corporate gain. Through our recently created company, Space 2 Inspire, we facilitate the introduction of a Brilliant Meetings culture in to the organisations we work with.

www.meetingexpert.co.uk has been created to document and share the best practices, information and templates used by the 'Brilliant Meetings' community and will be maintained and developed with all the passion and commitment applied to this book.

Acknowledgements

A 'thank-you' paragraph in the opening sequence to a book hardly justifies or rewards the care, enthusiasm and unwavering support, given freely and passionately by our fantastic families and friends. Particular thanks must go to our respective spouses, Elaine and Mark, for their patience and understanding during this process – thanks for believing in us.

Whilst we have received nuggets of knowledge and injections of information from all around us, specifically we want to thank and acknowledge Debbie Maitland and Hugh Davies. Debbie and Hugh selflessly reviewed and critiqued our developing manuscript, whilst offering insights from their specific areas of expertise and experience, which were invaluable to both us and the book.

Thanks to Liz Gooster at Pearson, for recognising the potential benefits of a new approach to business meetings, and for her support and belief during the evolution of *Brilliant Meetings*.

Duncan and Jane

Duncan and Jane can be contacted at www.space2inspire.co.uk

Welcome to *Brilliant Meetings*

You've probably picked up this book because you are continually frustrated by bad meeting experiences, and you want to begin a process for positive change. Like you, so many people – meeting participants irrespective of job level, culture or size of organisation – have low expectations of meetings simply because of repeated exposure to bad meetings.

If statistics were needed to prove how we feel about ineffective meetings, then research carried out in September 2008 on behalf of VisitBritain revealed that: 'Three in ten meetings are a waste of time, costing UK businesses a staggering £27.5 billion a year in wasted workforce hours'.

'The average working person spends eight working weeks per year in meetings – almost double a typical holiday allowance – and 29 per cent of these meetings were considered unproductive.'[1]

What is a meeting?

> It is a gathering of individuals collaborating with the interests of the organisation at its core, for a scheduled amount of time.

[1] This research, commissioned on behalf of VisitBritain, was carried out by Opinium Research LLP on an online poll of 2,226 British adults between 29 August and 2 September 2008.

What is a Brilliant Meeting?

It is the whole meeting experience; before during and after the physical or virtual gathering takes place, where the desired *learn*, *share*, *create* outcomes are achieved.

The recipe for a Brilliant Meeting.

All good recipes require the ingredients to be correctly procured and prepared, yet it is only when they are blended with the necessary knowledge and skills of the chef that the finished dish is a culinary success. In exactly the same way, participation, preparation and leadership are the ingredients that need combining in the right quantities to result in a Brilliant Meeting.

The ingredients for a Brilliant Meeting.

- Completed preparation
- Logistics – why, who, when, where, what
- Fully prepared agenda
- Ground rules – a common set of rules everyone abides by
- Start on time
- Focus on required outcomes
- Effective contributions and participation
- Presentations designed to stimulate and achieve meeting outcomes
- Finish on time
- 'Follow-up actions' assigned, understood and completed

The skills needed for a Brilliant Meeting

- The ability to 'choose your attitude'
- Construction and delivery of your 'Elevator Pitch'
- Making best use of available meeting resources and technologies
- Delivery of engaging presentations
- The correct use of questioning techniques

What Brilliant Meetings will do for you

- Produce fewer, shorter meetings
- Result in meetings that fully achieve their objectives
- Provide opportunities to showcase professional abilities and raise personal profiles
- Position meetings as an integral and important part of your job role
- Increase motivation through productive, effective meetings

How to use this *Brilliant Meetings* book

Brilliant Meetings is divided into four sections aimed at: Meeting Participants, Meeting Organisers, Meeting Leaders and the Organisation as a whole.

Each section is filled with templates, examples, tips, tricks and techniques aimed at improving the meeting experience.

The individual sections are easy to reference, providing you with exactly the information you need, when you need it.

How you can achieve a Brilliant Meeting

- Organising, participating and leading successful meetings are everyday requirements in business life, and they are within your reach if you follow the guidelines and advice offered in this book
- Promote the Brilliant Meetings ethos to your colleagues.
- Log on to www.meetingexpert.co.uk for continually updated material, templates, training, ideas etc.

Brilliant Meetings is a commonsense, practical and workable approach, designed to ensure that all meetings are:

- held for the right reasons;
- correctly prepared for;
- sufficiently documented;
- effective and productive.

Brilliant Meetings = more effective meetings

PART 1

Next time you participate in A Brilliant Meeting

Introduction

The most important people in meetings undoubtedly are the participants; the organiser has called the meeting to harvest their combined knowledge, experience and ideas, with the aim of achieving a successful outcome for the organisation. Yet frustratingly, most books that offer advice on meetings gloss over the presence of the participants, instead choosing to focus their content too heavily on the roles of the meeting organiser and leader.

How often have you heard or uttered any of the following immortal phrases, or sat passively in a meeting thinking:

- 'Yet another meeting about meetings.'
- 'What a waste of my time.'
- 'That did not achieve anything.'
- 'Now I can get back to my real job.'
- 'Death by PowerPoint™.'

Brilliant Meetings will equip participants with the knowledge and mindset to make real and valuable contributions to the meetings they attend. The ideas and inspiration necessary to transform participation in to a positive and motivational experience are presented in a format that is:

- easy to read;
- quick to reference for specific information;
- full of real-life examples;
- practical to implement.

Furthermore, the advice given is pertinent to everyday working life and, importantly, *Brilliant Meetings* illustrates how these skills can positively influence career opportunities.

A Brilliant Meeting requires participants with the right 'can-do' attitude. This will result in the collective efforts of the group being targeted correctly on producing the right decisions, which in turn positively drives the organisation forward, and makes working there more rewarding and motivating.

Let us now begin with item one on our agenda, with the ultimate goal of transforming your attendance into fulfilling participation.

CHAPTER 1

Using meetings to advance your career

Have you ever considered that meetings potentially offer you a brilliant opportunity to raise your profile and get yourself noticed as part of your everyday routine?

A meeting is a forum that provides you with a chance to create a persona that could impact on your ability to affect your own career, in either a positive or negative way.

In knowledge-based businesses, being constructive, confident and effective in meetings can be the difference between career success and stagnation. Demonstrating a willingness to share ideas and propose solutions to issues facing you and your department, will also illustrate that you have self-confidence and are engaged in what is going on around you.

Meetings are a good example of an area where we come into contact with people outside of our day-to-day regime, whether inside or outside of the organisation. Meetings can be complex because, unless they have been assembled solely for the purpose of putting out a leadership communication, it is almost impossible to predict what will be said, and who will say it.

There is a chance that, by the time you get to give your input, the meeting could already be running late, so how much time will you realistically have? The truth is that, long before you have your say, you will have portrayed valuable information about yourself; key subliminal signals about you and your capabilities that could help shape your career prospects, are already being sent out.

For the purpose of this chapter we are going to assume that meetings may involve people outside of your normal peer group, including people from outside of the organisation; people you are working in partnership with on a project, clients, or suppliers. There have been many examples of (new) careers being developed, as a result of positive exposure in meetings. Sometimes, without even knowing it, we find ourselves networking with new people within a meeting environment. By way of comment and feedback during a meeting we also have the opportunity to 'subliminally advertise' ourselves, which is much more effective and compelling than a photocopied CV that looks similar to everyone else's. Now is your chance to impress with your personality, experience and passion.

First impressions

We all know that first impressions are important; we are all guilty of judging others by their appearance, and so how you dress and your overall demeanour needs careful consideration, especially if you are not well known to everyone assembled. Whilst the collar and tie/business suit is no longer the only way to dress, and the concept of 'dress-down Friday' increases in popularity, how you appear will actually contribute to the way in which your words and deeds are interpreted by others. Armed with this knowledge, think carefully about the image you choose to portray through your appearance.

Advancing your career might mean an internal promotion, a move to another department, or even a move to a different organisation; perhaps even one of the other companies represented at your meeting. So, even before you have uttered a single word, you may have been judged already as being:

- convincing;
- trustworthy;
- professional;
- efficient;
- well prepared.

It is not just your appearance, clothes, hair, and the state of your shoes, etc., but the whole aura of your personality. Did you burst in a few moments late with a stack of papers and files under your arm, only to spend the next few minutes looking for the correct documents for this meeting? If you did, then you are on the back foot before you have made your first conscious contributions.

Choose your seat

If you have been given the choice of seats, does it matter where you locate yourself around the meeting table? Right from being an adolescent, in a theatre-style situation we avoid the front rows at all costs. In a boardroom style, if we know where the meeting leader is sitting, our first instinct may be to sit as far away as possible. The underlying fear is one of being put on the spot during the meeting without having a meaningful answer in response. If you want to get noticed and give your career prospects a boost, then sitting at the front or opposite the meeting leader is where you should be aiming for. By being fully prepared, there should not be any questions that you cannot handle, and you will be perceived as a confident, engaged participant right from the outset.

sitting at the front or opposite the meeting leader is where you should be aiming for

impact

- Arrive at the meeting early
- Claim your place around the table (preferably opposite the meeting leader)
- Initiate conversations with new members of the group
- Be prepared with questions and content

Your personal 'Elevator Pitch'

On the occasions when groups of people meet for the first time, quite often the meeting leader will give everyone an opportunity to introduce themselves. Typically people are truly unprepared for this very common occurrence and, generally, the introduction process then degenerates into one or two sentences from everyone that does little more than confirm their name, job title and length of time with the organisation. This format is usually then adopted by the remainder of the group, so little new information about the other participants is exchanged.

Sales people are schooled in the value of having an 'Elevator Pitch' (even if they are not familiar with the phrase) prepared about the product, service, or organisation they represent. When an opportunity presents itself to outline briefly what they are offering, they are ready with a concise 'hook' that will entice their listener to want to know more. The Elevator Pitch must be credible, succinct and must offer a solution to an actual business problem. A successful 'Elevator Pitch' – named after the amount of time you might get by chance with a key prospect who is unexpectedly riding in the same elevator – needs to be properly constructed

> The Elevator Pitch must be credible, succinct and must offer a solution

and practised to such a level that it sounds completely natural and credible. In the 'elevator' situation you have not got the time to go through your complete sales pitch, but if you can offer what sounds like the viable solution to a real business problem the prospect has, then there is a good chance that your 'hook' will give you the opportunity to sit down and present the whole sales pitch in due course.

brilliant definition

Elevator Pitch

Named after the amount of time you might get with a key prospect who is unexpectedly taking the same elevator, a successful Elevator Pitch needs to be properly constructed and practised to such a level that it sounds completely natural and credible.

So your personal 'Elevator Pitch' has to offer the same compelling insight into you, with the same hooks that leave your fellow meeting participants wanting to know more about you. It is the reason why you are employed by your organisation and are representing them in the meeting.

brilliant impact

Even if the meeting leader has not asked everyone to introduce themselves, think about giving a short personal introduction the first time you are asked to contribute.

'As there are some people here that I do not know, let me just briefly introduce myself . . .'

Remember that, whilst your delivery has to be natural (achieved through practice), it must also contain the following life saving **CPR** ingredients:

Concise

Professional

Relevant

'Good morning everyone. My name is Jane Hammersley. I have worked in the audio visual industry for over 10 years, in which time I have worked for resellers, distributors and manufacturers. More recently, with my business partner Duncan Peberdy, we have set up a new organisation called Space 2 Inspire, purely to focus on helping companies transform every meeting into a Brilliant Meeting.'

brilliant tip

A personal 'Elevator Pitch' is a long way from a CV, but in just a few words it tells those around the table that you have the credibility to be sitting in your seat and that you have had significant, relevant experience within the marketplace.

What is more, in the above example, your credibility has been reinforced, your personal confidence has increased, and you have gained the respect of the others around the table.

When there is a break in the meeting for refreshments, you have just given out some interesting information that should easily allow others to strike up a conversation with you.

Depending on the size of the meeting you might consider standing up to deliver your introduction. Even if those who preceded you remained in their seats, standing not only builds confidence and respect, but also ensures that everyone hears you clearly. The act of standing up will also encourage those present to give you their full attention.

If you have not heard of an Elevator Pitch before, there are plenty of good examples on the Internet with advice on how to construct them for maximum effect. Alternatively visit www.meetingexpert.co.uk for further guidance.

brilliant tips

1 Envisage how a guest speaker might be introduced at the annual event of a professional organisation. All their major achievements will be listed to justify the money you have spent buying a ticket to the function, with the draw of this big name being present. Use this as the starting point for collating your personal Elevator Pitch information, but tone it down and make it relevant to the subject of the meeting. Make it brief enough so that you do not come across as too arrogant.

2 Have more than one personal 'Elevator Pitch' prepared – think about the different content you would use at internal/external and formal/informal meetings.

Making the right contribution

Your energy level needs to be sufficiently high so that you project an impression of being motivated and upbeat, but not so much so that everyone thinks you have been consuming energy drinks all morning.

Pre-prepared notes and questions will increase your confidence

Whether you are delivering a formal presentation or just providing a verbal update on a project, prepare and visualise your opening words. Pre-prepared notes and questions will increase your confidence. Depending on the interaction during the meeting these might

tip

Prepare written notes containing your thoughts and ideas. Have questions detailed along with any supporting material. This will give your confidence a real boost.

have to change, but having them prepared will build your confidence, and a good start to the meeting will earn you respect.

Good communication is the key to your personal success

Good communication is the key to your personal success and, for you to demonstrate your capabilities – to help you stand out from your colleagues – you need to have total control over your 'performance'. Control means being fully prepared to present, whatever may conspire against you to reduce the impact of your message. In other words, you need to **PREEMPT** any eventuality that might occur.

Preparation

You have worked for hours on putting together a PowerPoint™ presentation, so what can go wrong? What if the projector is not working or the cable to connect it to your laptop is missing? How will that affect your confidence? But, instead of appearing disorganised and phased by the problems, if you know your presentation inside and out, and have the passion to deliver it without the planned resources, it still can be a rousing success.

Recap

If you want a key message to get through, then repeat it at least three times. It is no coincidence that advertisers repeat their messages during all the breaks on the same TV show. You can find many different ways to get the same message across, so your colleagues will not consciously recognise your winning tactic.

Educate

What is the point of delivering any contribution that does not contain new information? It is all very well being entertaining, even motivational, but to really make your mark focus on sharing information with the group that they did not already know. This will really set you apart, perhaps even single you out as a 'thought leader'; someone whose knowledge, experience and opinions matter.

Engage

Momentum

Given that so much of what we communicate is non-verbal, get up from behind the desk or lectern and fully interact with your group. Now your movement, coupled with your passion and knowledge, will produce a momentum to your presentation.

Passion

Passion is the element of your presentation that will stay with the group for the longest time. For your passion to be credible you must be well informed and knowledgeable; but above intelligence and accuracy, it will be your tone and passion for the subject that will be the most memorable.

Trust

Unless you make it very clear that you are making judgements, suppositions and estimates, the information that you present must be accurate. Handouts should expand on any statistics to prove their authenticity and, where necessary, supporting material should be available. Do not fall into the trap of presenting what you think people want to see and hear if it means 'bending the truth'.

In Chapter 4 we have provided comprehensive information and guidelines on developing and delivering presentations for anyone involved in meetings. When it is your own reputation on show, by being fully prepared for all challenging eventualities, you can prove to be more than just dependable. Now watch your career take off.

Using positive language

- Periodically reinforce the importance of your contribution by relating it to the purpose and objectives set for the meeting.
- Dependent on the meeting subject, use inspirational words to help your audience visualise your point of view: imagine, think about, consider, wouldn't it be fantastic if . . .
- Reaffirm 'your experiences and successes' to add value and credibility to your input.
- If you are going to aim a question at a specific individual, use their name at the start of the question so that they know immediately that this is directed solely at them.
- Avoid using jargon. Even if just one person does not follow what you are saying because of acronyms or technical phrases, and has not got the confidence to ask, then the impact is reduced.

brilliant tip

To assist you in remembering everyone's names, make a sketch of the table layout, and note where each person is sitting or arrange their business cards in order. After the meeting, this will make recall of people and their input more successful.

Meeting ground rules

Without a set of ground rules, participants will rely on their own interpretations of what constitutes acceptable behaviour in meetings. With the massive growth in wireless technology, two of the biggest contributors to interruptions are personal laptops and Blackberry™-style communication devices. Despite the door of the meeting room being closed, many meeting participants continue to do 'other work' with these devices. Some people even think it acceptable to answer their mobile phone, and then leave the meeting room to involve themselves in a conversation.

Every organisation will benefit from having ground rules in place that confirm to all what constitutes acceptable behaviour in meetings. These rules are suggested in Chapter 2. However, by adopting these guidelines in advance, you will earn the respect of your colleagues and hierarchy immediately. These are essential standards to ensure that meetings become as effective as possible, thereby increasing the overall productivity for organisations for whom meetings are the very lifeblood of product development, customer retention, improving business processes, etc.

Volunteer

Within many meetings there is often a necessary but unpopular job that needs a volunteer to take it on or otherwise needs delegating. Volunteering for this role – so long as you have the time and energy to make it a success and it is a project with high management visibility – will earn you some positive career credits in the short term. If you can deliver a successful outcome in the timescale and to budget, then the career points will multiply. Even if the outcome is not 100 per cent successful, your willingness to put yourself into the firing line by taking on such a job or task will not go unnoticed.

Likewise, if you have the opportunity to represent your organisation within a professional organisation or community project then take it on. You will then have two career advancing opportunities: as an official representative of the organisation and through increasing your network of valued contacts.

Conclusion

So, whilst meetings exist primarily to achieve the objectives of the organisation, never lose sight of the opportunities to further your own career by playing a full and active role in every aspect of them.

brilliant reminders

For using meetings to advance your career

- Look the part of an engaged and interested participant – remember first impressions really do count.
- Sit at the front, or opposite the meeting leader.
- Have a natural yet rehearsed personal 'Elevator Pitch' prepared.
- Show that you are listening attentively and actively by taking notes.
- Use positive language.
- Send an email to the meeting leader, thanking them for the meeting and giving any relevant feedback.

brilliant tip

If you do what you have always done, you will get what you always got!

CHAPTER 2

Meeting ground rules

Who needs more rules?

Just as young children thrive and flourish in an environment with clear and consistent expectations and boundaries, the success or failure of a meeting is also dependent on business guidelines and protocols that are clear, understandable and relevant to all. As participants, we all share responsibilities for both developing these guidelines and also adhering to them; in Part 4 the benefits of adopting these ground rules from an organisational viewpoint are outlined in more detail, but for now let us explore what differences we can make personally.

How often have you heard or uttered any of the following immortal phrases about meetings?

- 'Yet another meeting about meetings.'
- 'What a waste of time.'
- 'That did not achieve a single thing.'
- 'Death by meeting.'
- 'Now I can get back to my real job.'

Think back to the bad meetings that made you feel this way. Was it because the meeting started late, because too much time was devoted to recapping previous meetings, or because the scheduled meeting time had elapsed without conclusions being reached or actions assigned?

We know that all these and many more frustrations with meetings are an everyday occurrence as we experience them for ourselves, and survey after survey confirms this.

'Most professionals who meet on a regular basis admit that they do the following: daydream 91 per cent, miss meetings 96 per cent, miss parts of meetings 95 per cent, bring other work to meetings 73 per cent.'

Survey by MCI Conferencing

'Workers spend an average of 5.6 hours per week in meetings and 69 per cent of responders felt that meetings are not productive.'

Microsoft survey tracking office productivity around the world

'Biscuits are the key to a successful meeting', according to a survey of 1,000 professionals. 80 per cent of respondents believe that the better the biscuit presented, the more successful the meeting.

Poll by Holiday Inn

'49 per cent of participants considered unfocused meetings and projects as the biggest workplace time waster and the primary reason for unproductive workdays.'

An article in the Autumn 2006 issue of The Facilitator Newsletter

If you want to take an active role in transforming those 'waste of my time' meetings into productive and inspiring experiences, we challenge you to adopt the following **ACTION PLAN** for yourself, and ask your fellow participants to step up to the mark.

The Brilliant Meetings ACTION PLAN

Arrive on time, with relevant, well-prepared content
Choose your attitude (see Chapter 5 for more details)
Turn off all personal communication devices: phone, Blackberry™, laptop, etc.
Imagine your Chief Executive Officer (CEO) or Managing Director (MD) is present
Obey the agenda and stay until the end of the meeting
Never use jargon and avoid distracting side conversations

Participate actively – 'silence is acceptance'
Learn what you do not know, share what you do know
Accept and fully support consensus decisions
Named actions must be completed

Let us take a look at each of these in turn.

Arrive on time, with relevant, well-prepared content

Your prompt arrival and preparation impacts on more people than just yourself – respect other people's time as a valuable resource.

respect other people's time as a valuable resource

example

No one ever got fired for arriving early.

You are not in the meeting to make up numbers, so if you are late the meeting leader has a tough call to make. Postpone the start – thereby wasting the time of everyone else in the meeting and reducing the overall meeting time – or to proceed without you – thereby excluding your potentially valuable input. The same is also true when the meeting is late reconvening after a break.

brilliant tip

Build in a buffer time for unexpected travel situations – aim to be there early. If you are already fully prepared, use any spare time effectively catching up on work, phone calls, networking, etc. Being early will also allow you to 'choose your seat' and being well prepared affords you the comfort of taking a position opposite the meeting leader and to be fully engaged in the meeting right from the start. (See Chapter 1.)

If the meeting you are scheduled to attend is a follow-on meeting, ensure that all of your previous actions have been completed. Prepare to brief the group on the results of your actions with either a formal presentation, verbal update, or a handout.

If you are presenting material, develop your presentation in a natural format for *you*. It does not have to be PowerPoint™, you can use handout sheets, a flip chart or dry erase board. All of these methods can be effective, as long as you spend time preparing all relevant material and rehearsing the delivery. For more suggestions see Chapter 4, Contributions and presentations.

If, during the meeting, you need to reference background material, ensure that this is circulated well in advance of the meeting.

Try to 'second guess' any questions that may arise as a result of your contribution and develop possible responses to these.

'Before anything else, preparation is the key to success.'

Alexander Graham Bell

Choose your attitude

The ability to choose your attitude – is the glass half full or half empty? – is incredibly important to your participation in the meeting, and therefore the overall outcome. If, as you sit down, you have already determined it to be a waste of your time, then it probably will end up being just so.

However, if you believe that your contribution and participation will help to transform a potential 'waste of time' meeting into a productive, effective and inspiring experience for all, then choose your attitude accordingly and give the meeting a chance. This is expanded upon in Chapter 5.

If you follow our guidelines in Chapter 3 and accept the meeting invitation, you must then recognise the meeting as an essential part of your role, and not something that 'takes you away from your job'.

Turn off all personal communication devices

Conference phones should be the only phones used in a meeting environment. They are a group device and should be used when all those present need to participate in the same conversation together.

Personal mobile phones should be turned off before you enter the meeting room. If it is important enough for you to be in the meeting, then it is important that you and the other participants are not interrupted. There will of course be exceptions to this rule: an expectant father, a production problem, financial issues, etc., and these will be handled on a case-by-case basis by the meeting leader.

brilliant tip

It costs little or nothing to change your voicemail to let callers know, for example, that you are in a meeting until 2pm. This sets the expectation that they will not receive a call back until after this time, and alerts them not to call you again. The same is true with an automatic 'out of office' email responder message which details your availability.

Even those participants who have purposefully placed their phone on the desk in 'silent' mode will now be distracted by the same caller repeatedly trying to get hold of them. They may also be tempted to type a 'quick' text message or reply to an incoming text or Blackberry style email.

The culture around laptops in meetings requires a special mention as their use is really a symptom of a poor meeting rather than the cause. When laptops were still tethered to network cables, there was little else one could do with them in a meeting except take notes, or present with PowerPoint™. With the increase in wireless communications, on the pretext of taking notes, the reality is that email or other work is being undertaken.

Even if you are just taking notes, to the presenter in the meeting you are giving the impression of not fully listening, which can lead to resentment, but for most of us it probably means we are not paying full attention to either the email or the speaker alone. Now we are doing two jobs poorly instead of concentrating all our efforts on just one.

Imagine your CEO or MD is present

The truth is that when the CEO, MD or senior board members are present we do act differently, even where standards have not been published. We will not leave our phones on, we will not talk across someone else, we will be fully prepared, and we will be sure to arrive early. Because we know the CEO or MD will be sitting across the table we may also enter the meeting with a different mindset. This meeting now matters, and we have arrived with the right attitude that will see us making a difference.

This is exactly the mindset – which you can absolutely choose – that you should adopt for every single meeting.

Obey the agenda and stay until the end of the meeting

In general do not raise any issues that are not on the agenda unless they have become critical to the meeting. Even then, it should be the decision of the meeting leader to allow any deviation. Agenda quite literally means 'things that have to be done', a list of discussion/presentation topics drawn up by the meeting organiser in order to achieve the intended outcome within the allotted time.

> Agenda quite literally means 'things that have to be done'

Too frequently participants try to raise other issues as a deliberate spoiling tactic, especially if the decision being made does not suit them personally.

If you have been given a presentation slot in the meeting agenda, ensure that you keep the content relevant as well as keeping to your allotted time, remembering to factor in enough time for questions.

Never use jargon and avoid distracting side conversations

Avoid using jargon. Even if just one person does not follow what is being said, the impact of your input is reduced and you have left them feeling uncomfortable with their perceived lack of knowledge. When giving technical information, always pitch your presentation at the participant with the lowest perceived

> always pitch your presentation at the participant with the lowest perceived subject knowledge

subject knowledge, taking care to explain fully any 'buzz words', acronyms or technical terms.

Having private conversations within a meeting promotes a very negative situation: those engaged in the conversation are not listening to the other participants fully. The other participants can easily make two assumptions. First, because the comments are not being made to the whole of the group, they must in some way be divisive. Second, that their contribution is not worthy of everyone's full attention. This situation promotes distrust and should be avoided. If this situation arises during a presentation that you are delivering, either pause, waiting for the conversations to cease, or alternatively ask the participants to share their views with you and the rest of the group.

Participate actively – 'silence is acceptance'

You must ask questions when anything is unclear, and usually the presenter will have announced at the outset how and when questions will be dealt with. Even if they have indicated that interruptions are welcome, they might ask you to 'hold that thought' until later. If you have a problem with any of the contributions or viewpoints being stated, you must bring it up during the meeting so that the group can hear your reservations and discuss accordingly.

Making notes also clearly illustrates that you are fully engaged, that you are taking the meeting seriously, and that you are actively listening and participating. If there is a designated note-taker, or pre-prepared notes are available as a handout, there is then a temptation to rely on those sources and not to take notes of your own. The trouble with this approach is twofold. Firstly, those notes might not contain the things that were of importance to you; after all, they are notes and not a verbatim transcript.

Secondly, taking notes allows you to write down anything that is unclear to you, so that you can raise it at the appropriate moment.

Learn what you do not know, share what you do know

brilliant impact

Consider every meeting as a great opportunity to learn what you need to, whilst simultaneously sharing what you know. This will increase knowledge transfer within the meeting, and add to its success.

Accept and fully support consensus decisions

There are many ways of reaching decisions and, whilst one based on consensus of opinion can take longer to achieve, it keeps groups on track and together, rather than creating divisions through an 'either or' vote. Consensus does not necessarily mean that all individuals think that the decision made is the best one possible or even that they are sure the decision reached will work. What it does mean is that, in coming to that decision, no one felt that their own position on the matter was misunderstood or that it was not given a proper hearing. The hope is that everyone together will think it is the best decision. In a process looking for a consensus decision, either the group achieves the required outcome, or no one does.

Remember, meetings are not social events; they are where business decisions are advanced primarily for the benefit of the organisation, not for the empire building of individuals or individual departments.

Named actions must be completed

Meeting actions resulting from your Brilliant Meeting should now be completed.

During the meeting, ensure complete understanding of any actions that you are partly or wholly responsible for. This should include how the action should be reported on and what timescales are attached to it, whilst making sure that you have the right resources in place to achieve the desired outcome.

Conclusion – ACTION PLAN

By adopting the **ACTION PLAN** you will enhance your meeting effectiveness and quickly earn the respect of your colleagues. Respect is hard to gain and easy to destroy, but essentially if you choose to be punctual, participate in a positive manner and complete your actions, your colleagues will naturally award you the respect you deserve.

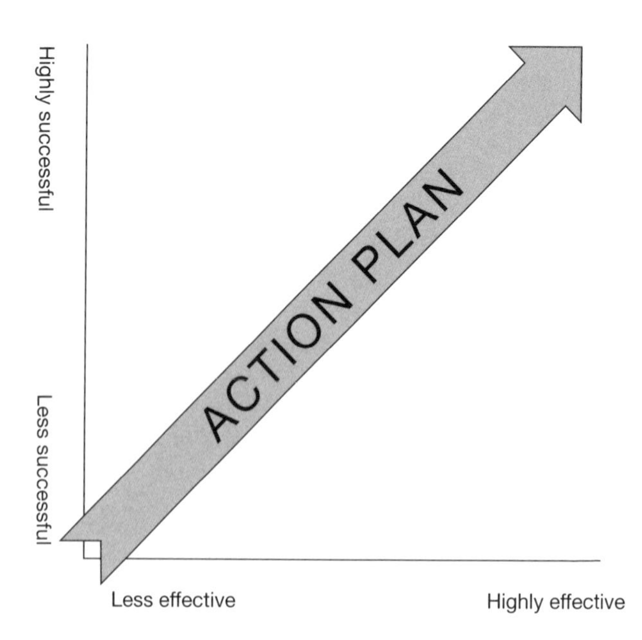

CHAPTER 3

Before the meeting

Every meeting starts long before you walk through the meeting room door.

> Every meeting starts long before you walk through the meeting room door

If it is a series of meetings then you may have previous assigned actions that need completing prior to the next meeting. If it is a new meeting then you should have received the agenda and seen the list of participants.

As a participant, it is your responsibility to bring as much of your experience and knowledge into each meeting that you attend. Why else have you been invited?

Every participant must have the mindset that pre-meeting preparation is essential to complete; not just for themselves, but for the other participants as well. For example, it is not just your reputation that is damaged if you have failed to assemble all the information requested of you, but what about the reduced value to your fellow participants? Just like you, these participants may have travelled extensively or rescheduled other tasks in order to be there.

So as soon as the meeting invitation arrives in your in-box, begin to mentally calculate the necessary time and resources needed to complete the meeting preparation, and what impact this has on your schedule. How much more could be achieved by everyone with thorough preparation?

Meeting request, meeting invitation, or meeting demand?

Most of us receive our instruction to participate in a meeting electronically. On the screen is the name of the organiser, the date, time and venue, together with information to let us know if we are 'Required' or 'Optional'. Our electronic diary will immediately warn us if there is a conflict with an existing entry in our calendar, and buttons appear prompting us to 'Accept', 'Decline', or to 'Propose a New Time'. With a couple of nonchalant clicks of our mouse the time is allocated in our calendar.

Electronic communication is invasive although we have become conditioned to give emails and telephone calls immediate priority. Emails that pop up onto the screens of our computers and mobile phones take away our attention – however momentarily – from what we are doing. Too often we respond on impulse; we deal with the intrusion – in this case we accept as there is no scheduling conflict in our diary – and carry on. We have taken the easy option, the route of least resistance.

Every time a meeting is requested from now on you must consider the implications on yourself, the other meeting participants, and then determine what is best for you and your organisation. Just because meetings within your organisation have always occurred this way, does not make it right and certainly should not mean that they continue unchallenged and unchanged.

Only accept if:

- in your opinion a physical meeting is the right way to deal with the issue;
- by attending you can add real value to the meeting;
- you have time to assemble all the necessary information and prepare correctly;

- the desired outcome can be achieved or advanced significantly;
- the date, time and venue is convenient;
- you fully understand the objectives, decision-making process, etc.

So, once you have carefully considered your personal value to the meeting, if you think that your attendance is not a worthwhile use of your time, communicate this back to the organiser as soon as possible. Do not just decline, but outline the reason(s) why. It could be that you have raised a point that the meeting organiser has overlooked, and the sooner they are aware of it, the sooner any decision about cancelling or rearranging the meeting can be made and communicated to the other participants.

Thank the organiser for sending you the invitation, and then outline your position.

Review the agenda

The word agenda is derived from the Latin as a plural of agendum and, quite literally, means 'things to be done'.

Look at the agenda with the same focus that you applied to the meeting invitation.

- Are the objectives fully supported by the agenda items?
- Can the agenda be effectively completed in the time allocated?
- Have all the necessary topics to 'gain an advance' been included?
- Will all necessary actions assigned to you at the previous meeting have been completed?
- Is there the authority within the meeting to reach binding decisions?

If the answer to any one of the above questions is 'no', then the meeting should not go ahead in its proposed format. Bringing people together costs far more than their time alone; the property costs for meeting facilities, together with any additional travel, accommodation, and refreshments are considerable. If the meeting cannot hope to meet its outcomes then it is better it did not take place. If it were to proceed, what other outcome could there be apart from the well clichéd 'meeting about a meeting?' In other words, this can only be at best an ineffective meeting. Ineffective meetings trigger low morale, low productivity and, if not actively challenged, can easily become the norm, dragging down standards within organisations.

Whilst it can be argued that a meeting organiser should not circulate an agenda that has no realistic chance of being completed during the meeting, it is also the responsibility of any participant to notify the organiser if they feel the intended outcome(s) is unachievable. The participant may have information not available to the organiser, and so a culture must persist that allows such honest feedback.

In order to react to issues and situations that arise and have to be dealt with almost immediately, unplanned or impromptu meetings are an essential part of everyday business life. However, every planned meeting should have an agenda, without exception.

every planned meeting should have an agenda, without exception

Considering your contributions

Having looked carefully through the agenda, in addition to any presentations you have to prepare for, you will also be able to identify any discussions that may require your input. Given that you can see the topics being presented by others, is there anything that you can research or prepare in advance that will support the position you are most likely to endorse?

Obviously, the sooner you start your research and identify the questions you could ask, etc., the more background material you will be able to pull together in full readiness.

For your own planned presentations, the content, design and delivery are all key to persuading others to take your point of view.

Chapter 4 is devoted to planning, preparing and presenting during a meeting, taking into consideration the resources and equipment available to you.

brilliant reminders

Before the meeting takes place do the following.

- Review the agenda. Do you need any information from colleagues for this meeting?
- Review your authority with regard to making decisions on the intended outcomes.
- Raise any concerns or required clarifications in advance of the meeting.
- Prepare and circulate any briefing notes.
- Plan to arrive early.

brilliant tip

All too often we determine the required travelling time by what has been our best experience to date. Now we have no 'buffer' built in if something goes wrong. Bad weather, traffic accidents and congestion are all out of our control, and are almost certain to conspire against us when we have not left sufficient time.

Aim to be there early and, if you are fully prepared on arrival, use the time to catch up on telephone calls.

CHAPTER 4

Contributions and presentations

Whether your role in the meeting is to participate, prepare or lead, there may be occasions when it will be necessary to deliver valuable contributions. Presenting your views and recommendations requires preparation, communication skills and personal confidence in order to get your message across clearly.

brilliant tip

Brilliant ideas are not recognised until they are effectively communicated.

The contribution of ideas, facts, opinions, figures etc., can be delivered in many ways, depending on, amongst other things, the formality of the meeting, what resources are available and what information needs to be imparted, but, whatever the method, preparation is essential.

preparation is essential

Microsoft PowerPoint™, together with a projector, is generally the solution of choice to deliver a presentation to a business meeting, and this will be covered in more detail later in this chapter, but let us now begin with the preparation phase.

Planning your contribution

A full understanding of the purpose of the meeting, along with your role within it, is crucial to developing effective contributions. It is also essential that you establish the needs and expected outcomes of your audience. The next step is to prepare your contributions. You may have been asked to contribute in some form or other already, or perhaps you are preparing yourself 'just in case' your opinion is sought.

Starting a presentation from scratch can often be a daunting task, especially if you are beginning with a 'blank piece of paper' in the hope of transforming it in to a memorable inspiring presentation, in a matter of minutes. But help is at hand.

Purpose

First, write down the purpose of your contribution/presentation; this will help to start the process of getting your message across in a clear and effective manner. The purpose can either be sketched on paper, or typed in to your computer in outline form.

brilliant tip

Do not fall in to the trap of typing content straight in to PowerPoint™ slides before the entire contribution/presentation has been outlined.

List

Once you have decided on the main purpose of your message, now begin to outline the information you want to share. Form a list of all the information needed and collate all relevant material before you begin to piece them together. Once your content is gathered, group the information by subject and organise it in order of importance relating to the purpose of your contribution. This will help to ensure that information is in a logical sequence and is easily understood.

Detail

Note down all the points that you want to get across in each section, and the look and feel of your presentation will start to take shape.

Structuring your contribution

tip

Repetition = Retention

Now that the content for your contribution has been outlined and collated, fundamentally there are only three things to remember:

- **Tell them what you are going to tell them.**
- **Tell them.**
- **Tell them what you told them.**

This method allows you to reinforce your message at least 3 times, making it more likely that your content will be listened to and remembered for longer than 20 minutes!

brilliant example

It has been suggested that, if you make a point only once, at the end of your presentation, just 10 per cent of the audience will remember it. If you repeat a point six times, retention jumps to 90 per cent. Without repetition, 40 per cent of your audience will forget virtually everything you said within 20 minutes of your conclusion. Within 24 hours, 70 per cent of the audience will forget almost 100 per cent of your message.

Let us look at the speech made by Sir Winston Churchill in June 1940, 'We Shall Fight on the Beaches'. Although this is a fairly lengthy speech, in the final paragraph, which lasted no more than one minute, he says, 'We shall fight . . .' SEVEN times and 'We shall . . .' ELEVEN times.

Think about how you will start to draft your presentation – do you write it in the order in which you will deliver the presentation, from the beginning, and slog through the content to reach your conclusions? There is a better way to avoid procrastinating about the preparation because of the perceived enormity of the task.

Begin with – The conclusion

Tell them what you told them.

Starting with the end in mind, focus on the purpose of your presentation, relating to the objectives of the meeting, whilst keeping your audience in the forefront of your mind. Summarise succinctly the main points of your presentation, this will help to 'fix' them in the memories of your audience. This is what you want your colleagues to 'take away' from your presentation.

> Always conclude the presentation with a strong positive statement

Always conclude the presentation with a strong positive statement, remember the words of Sir Winston Churchill that were repeated 11 times in less than a minute, 'We shall'.

You now have the opportunity to define your destination – you know where your presentation is heading, and your audience will understand fully your message.

Develop – The introduction – A strong opening message

Tell them what you are going to tell them.

Next, develop the attention-grabbing opening to your presentation that concisely spells out the benefits of engaging with your presentation and clearly articulates what your presentation will cover.

Cast your mind back to Chapter 1 – Using meetings to advance your career – you have developed and practised your personal 'Elevator Pitch' – now could be the time to put it in to action. The first two to three minutes are crucial for you to confirm your credibility and engage your audience. If you are presenting to a group that does not know you, you may choose to open with your Elevator Pitch, alternatively you can launch straight in to your well-practised, professional presentation. Lead with a rhetorical open question then follow with compelling statistics or paint a vision of how things could be for them, to tweak their curiosity.

The first two to three minutes are crucial for you to confirm your credibility and engage your audience

Inform your audience at the outset how you will deal with questions. Will you deal with them as they arise, or do you intend to 'park' questions by noting them on a nearby flipchart, then cover them at the end? Allow your audience to seek clarification throughout where necessary, and define at the outset if there are any scheduled breaks planned for your session.

brilliant tip

Interaction

At the beginning of your presentation ask a question that prompts a response with a show of hands. This will give you three main benefits:

1. You can quickly gauge group opinion and/or understanding.
2. The group is now actively engaged with your topic.
3. It demonstrates your interest in group opinion.

And finally – The middle

Tell them . . .

The middle should now be much easier to write as essentially you have summarised it twice already in the introduction and conclusion. Use information only in which you are 100 per cent confident and ensure that you have thorough knowledge of the subject. Make the content as 'conversational' as possible, try to avoid very formal/stiff presentations. Use examples where possible to illustrate your point, making it relevant to your audience.

Whatever the ultimate method of delivery for your contribution (PowerPoint™, flip chart, verbal update), always try to vary how the information is presented to your audience, using a mixture of text, graphics, images and verbal reinforcement.

brilliant example

We retain 10 per cent of what we read.

We retain 20 per cent of what we hear.

We retain 30 per cent of what we see.

We retain 50 per cent of what we hear and see.

We retain 70 per cent of what we say.

We retain 90 per cent of what we do.

These statistics illustrate the need to reiterate and repeat important messages as *we only retain 50 per cent of what we hear and see.*

brilliant reminders

To give yourself the best possible chance to capture and maintain the attention of your audience as well as being memorable (for the right reasons!), when planning your presentation keep in mind these simple ABCs:

- **A**ppealing – in terms of content and delivery;
- **B**elievable – everything should be factual and deliverable – not being able to deliver something you have personally promised will not be good for your career. Always have research, statistics, expert opinion or other evidence available to reinforce your position.
- **C**lear – see information on the use of positive language as detailed in Chapter 1, being especially mindful to avoid jargon.

Contribution/presentation delivery options

Having drafted your contribution, check what equipment is available to present it during the meeting. The most commonly used meeting resources are flip chart, dry erase boards and PowerPoint™ delivered through a projector. Certainly Microsoft PowerPoint™ is the most widely used business presentation tool and, consequently, this will be covered in much more detail. But if this option is not available, or you would prefer to use something else, here are some Brilliant Tips for using other meeting room resources to deliver your contribution/presentation.

Flip chart

This is a group 'tool', great for real-time collaboration and audience participation. Before the meeting begins check that there is enough paper left on the flip chart for your presentation and that there are marker pens available that work!

brilliant tip

Make the first flip chart page a title page to include the subject matter and your name. This ensures that the body of your presentation is kept hidden until you are ready.

You can prepare flip charts in advance of the meeting, noting down the main points that you plan to cover and asking for feedback which you can then record. Preparing flip charts in this way can save a considerable amount of time during your presentation.

brilliant tip

Use a pencil to add reminder notes at the top of each flip chart page in advance of the meeting. You will be able to see your notes close up, but they will be invisible to your audience.

Leave a blank page in between each prepared page for additional notes.

If you are using multiple flip chart pages, explore the possibility of 'posting' these around the room, thereby giving everyone a constant view of your presentation. Some meeting rooms are equipped with rails and magnets for this purpose, and we would strongly advise you to use this facility if it is available.

brilliant tip

Ensure that your flip chart is not overloaded with text and use only the top 2/3 of the page, so that even people sitting at the back can view all of the information.

Have a conclusion page at the end. This acts as a great reminder to summarise what you have said, therefore helping the audience to remember your messages.

Repetition = Retention

Dry erase boards

Seek permission to enter the meeting room in advance in order to 'write up' your presentation, in the same way as teachers did in their classrooms 'in the old days'!

brilliant tip

Dry erase boards are great for sketching out ideas and drawings – use dark, bold colours for text and vibrant primary colours for drawings.

Write up only bullet points and complete the rest whilst you are delivering your presentation. Remember to get someone to copy the finished information from the board at the end, or better still take a photograph of it, transfer it in to a jpeg file and include it in the meeting notes.

brilliant tip

If you have a magnetic whiteboard in the room, prepare your content on coloured paper in advance and 'post it up' during your presentation for dramatic visual effect.

Microsoft PowerPoint™

The visual aid of choice for providing information during meetings is Microsoft PowerPoint™. Like the majority of computer programs, most of us use only around 10–20 per cent of their capabilities; however, the problem with using such a highly visual and visible tool as Microsoft PowerPoint™ is that we publicly share our lack of knowledge, potentially without realising it.

We have all been in meetings where the information is presented from PowerPoint™ through a projector or a plasma screen. The last thing an audience wants to see is reams of information on a screen, which the presenter then proceeds to read out word for word. It happens, and all too often. The other way to show how little you really know about PowerPoint™, is to have words that move in with some sort of animation, perhaps accompanied by a crass sound, and then presented in colours that contrast poorly.

The effect of these inadequacies is commonly known as 'Death by PowerPoint™' – a sure way to lose your audience and with it your credibility.

To guard against finding yourself in this 'nightmare' scenario, when preparing a presentation we strongly recommend following these general guidelines:

- Use PowerPoint™ *only* as an "Aide Memoir", thoroughly preparing and practising your material so that you can deliver the content relevant to each slide heading in an engaging and informative way.
- Remember the KISS principle (Keep It Simple, Stupid).
- Use graphics and pictures where possible, with any text in uniform fonts and sizes whilst ensuring that it is large enough for everyone to read.

- Limit the amount of information per slide; a maximum of five bullet points.
- If your organisation does not have a standard template for PowerPoint™ presentations, then select colours that have the right contrast and brightness when projected or shown through a plasma or LCD screen – dark text on a light background is best, but avoid white backgrounds – tone it down by using beige or another light colour that will be easy on the eyes. Maintain the consistency of your colour scheme throughout your presentation.
- Learn how to link to additional information or to another PowerPoint™ file from a slide in your master presentation so you can switch seamlessly to this additional information without losing the concentration of your audience.
- Always remember that PowerPoint™, however well presented, only equals information, whereas demonstrating knowledge is something much more powerful altogether.

Anybody who can present information effectively is going to be of more value to an organisation than somebody who cannot. Two people may have the same technical knowledge and experience, but the one that can engage better with clients, suppliers and colleagues, will influence a different, more positive outcome.

Choose the right attributes for your slides

For the majority of PowerPoint™ presentations the following guidelines are easy to remember and should prevent you from succumbing to the temptation of 'bullet point' overload and allowing your audience to read the information before you have had a chance to!

brilliant tip

Use the 10/20/30 rule for PowerPoint™ presentations:

- 10 slides maximum;
- 20 minutes maximum;
- 30 point font size as the minimum.

Now you cannot have large paragraphs that are read out verbatim.

Rule developed by Guy Kawasaki – www.guykawasaki.com

Slide design

Now is the time to begin transferring all of your content on to the slides, but *before* you commence, set up a slide template using the 'Slide Master' to determine the design of all of your slides.

brilliant tip

Use the Slide Master option in PowerPoint™ to create a uniform style for your whole presentation. This will save you hours of extra work formatting every single slide.

Insert your organisation's logo in to the Slide Master, instead of placing it on every single slide, along with version information, and *your name*. This is also the place to set up fonts, size, and slide colour scheme.

brilliant tip

If you like the design and layout of someone else's PowerPoint™ presentation and have modified it to promote your own content, do not forget to change the title, author and organisation details in the summary dialogue box, under the presentation properties. (File, Properties, Summary.)

Title

Make your slides easy to follow. Put the *title* at the top of the slide where your audience expects to find it. Phrases should read left to right and top to bottom whilst important information is detailed near to the top of the slide. Often the bottom portions of slides cannot be seen from the back rows because heads are in the way.

brilliant tip

Write an 'attention getting' title for each slide to spark the interest and curiosity of your audience.

Font

Choose a font that is simple and easy to read such as Arial, Times New Roman or Verdana. Avoid script-style fonts as they are hard to read on screen. Use no more than two different fonts – one for headings and another for content. Keep all fonts large enough (at least 24 pt and preferably 30 pt) so that people at the back of the room will be able to read easily what is on the screen.

Colour

Use contrasting colours for text and background. Dark text on a light background is best, but avoid white backgrounds – tone it down by using beige or another light colour that will be easy on the eye. Dark backgrounds are effective to show off organisational colours or if you just want to dazzle the crowd. In that case, be sure to choose a light colour for the text to ensure it can be read easily and be aware that patterned or textured backgrounds can reduce readability. Maintain the consistency of your colour scheme throughout your presentation.

Transitions and animations

Avoid excessive use of slide transitions and animations. Whilst transitions and animations can add drama and build anticipation, too much can be distracting and confusing. Remember, the slide show is meant to be a visual aid, *you* are the focus of the presentation. Keep animations consistent in the presentation by using animation schemes and apply the same transition throughout.

Bullet points

Simplify the content by giving a key message on each bullet point that you can expand on during the presentation. Make the information on the bullet point as short and as 'cryptic' as possible, that way your audience will not already know what you are going to say. Keep the most important points near the top of the slide for easy reading in the back rows. Focus on one topic area and use no more than five concise bullets per slide.

Non-text content

Try to make the slides visually appealing by combining photographs, charts and graphs with the text. Avoid having text-only slides that your audience can read before you have the opportunity to add your comments verbally.

Slide order

Slide order can be changed easily after the information has been prepared. Do not get too 'hung up' on the order until you are ready to begin your rehearsal, and then you can manipulate the sequence of your slides through the 'slide sorter' option.

Hide your options

tip

Prepare additional slides to cover any questions that you think *may* arise during your presentation and access these at any time.

Ensuring that you stick to time is essential, but so too is ensuring that you are sufficiently equipped to deal with those predictably awkward questions. There are essentially two ways in which you can pre-prepare extra slides that you do not plan to use in the main body of your presentation. They can be accessed at any time.

1. Prepare your slide in the normal way then right-click on a slide in normal view and choose 'Hide Slide' – it will not then display during your standard presentation. If you want to show a hidden slide during the course of your presentation, right-click the mouse, select 'Go' and then 'Slide Navigator'.
2. Place a 'hidden' button on your main slide(s). Insert a hyperlink to an additional slide that contains your 'extra' information. You can then choose when, or if, to use the option to view additional hidden slides during your presentation.

Using hidden slides and hyperlinks allows you to tailor a presentation to suit a particular audience by covering more specific subjects that are not necessarily of general interest.

brilliant summary

Preparing a presentation using PowerPoint™

Slide Master – use to maintain consistency throughout presentation.

Limit use of different fonts, colour, animation and transitions.

Information should be kept to a minimum and presented with a mixture of bullets and graphics.

Death by PowerPoint™ should be avoided at all costs and keep to your allotted time!

Establish your credibility, talk confidently and knowledgeably about your subject.

For more great tips on how to use PowerPoint™ to great effect visit www.meetingexpert.co.uk

Handouts

Delivering a fascinating presentation that inspires your audience is all very well, but memories fade and you are not always on hand to add your verbal content, wit and repartee! To reinforce your presentation, you should think seriously about providing handouts for your audience to take away. It is not good enough simply to print out your slides and give them out; you should consider adding notes to these slides in PowerPoint™ and printing out the notes pages in advance of your presentation. That way you will also have notes that you can refer to throughout the presentation.

brilliant tip

Handouts allow people to concentrate on you, and not on taking notes. Ensure any handouts have your name and contact details on them, preferably as part of the 'header/footer' information. If you are going to hand these out at the beginning of your presentation to allow your audience to make any additional notes, ask that they refrain from reading the notes ahead of you.

If you have notes to hand out after the presentation, inform your audience of this at the outset.

Delivering content

Remind yourself that your presentation is for the benefit of the audience – a rapid-fire list of bullet points, read verbatim from a screen/flip chart or whiteboard, is less than inspiring and of little benefit to them. Your visual aids may have no text at all – the meaning, content and context is delivered by you. The slides are there to support you, not to be read more quickly by the audience than you can recite them!

Body language

Body language is a significant indicator of your work ethic, enthusiasm and attitude. More than the words you use or the tone of your voice, your physical behaviour communicates everything to the keen observer. Reputations (good or otherwise) are quickly gained from 'first impressions', how you look, act and react – body language.

There are many different scenarios that demand different types and styles of body language in order to make a positive impression on clients, peers and management. Some circumstances are more formal than others, but overall you need to understand how your own body language will be interpreted, and how to interpret that in others.

---

brilliant reminders

Presenter's body language

- Speak in a clear, confident voice that can be heard by everyone.
- Walk around to engage your audience, but avoid excessive movement.
- Make eye contact with everyone in your audience.
- Touch, turn, talk. If you are using a visual aid as part of your presentation, do not be tempted to talk whilst facing the slides or flip chart.

Keep to time

There is nothing worse than a presenter over-running as a result of a long boring presentation. Ensure that you have built in enough time for questions and feedback, finishing ahead of your allocated time – leave them wanting more from you.

Do not hide

Now is your chance to shine, do not hide behind a lectern or table. If possible move to one side and physically engage with your audience, possibly walking around the room to capture attention. Always stand up to make a presentation, your voice will be louder and directed at your audience which will in turn capture their attention and give you more confidence.

Practise, practise, practise

tip

When actually delivering your contribution/presentation, irrespective of the delivery medium, remember the five **C**s for delivering Brilliant content:

- **C**onfident;
- **C**redible;
- **C**ompetent;
- **C**onvincing;
- **C**omfortable.

brilliant reminders

Delivering content

- **O**utline your objectives at the beginning of the presentation (keep them visible throughout if possible).
- **M**ake eye contact with everyone in the meeting.
- **K**eep your hands steady and avoid fidgeting.
- **S**peak to the participants, not to the visual aids.
- **S**peak in a clear, confident voice that can be heard by everyone.
- **S**peak with intonation and rhythm in your voice to emphasise your main points.
- **K**eep to your allocated time.
- **U**se the KILL principle when using a flip chart – **K**eep **I**t **L**arge and **L**egible.

brilliant tip

For much more insight in to presentations, invest in *Brilliant Presentation* by Richard Hall. www.pearson-books.com

Conclusion

brilliant tip

Repetition = Retention

To be a good presenter you need to be engaged with your audience and know your topic inside and out. Keep the presentation concise and include only relevant information. When using an

Keep the presentation concise and include only relevant information

audio visual aid, such as PowerPoint™, use *only* as an accompaniment to your presentation to reinforce your point, not as a crutch.

Remember – your slide show is not the presentation – *you are the presentation.*

The final task is to practise your presentation on your own. This will allow you to become very familiar with the content, subject and also confirm how long it will take you to deliver. When you are happy with it, try it out on a friend. Now you are ready to really 'make your mark'.

Speak to your audience, not to the screen.

brilliant resource

For great presentations and PowerPoint™ resources visit www.meetingexpert.co.uk.

Checklist - Brilliant Contributions & Presentations

Summary Information
Meeting Title
Name of Group
Date & Time of Meeting
Meeting Venue
Presentation Topic

space2inspire

	Done	Comments	Date/Ref
Content And Presentation Completed			
Presentation Rehearsed, And Fits In To Allocated Time			
Speaker Notes And Handouts Prepared			
Prepare Any Resources; Flipchart, Dry Erase Board Etc.			
Check Resources To Be Used To Deliver Presentation			
Computer Presentation Saved On To Computer And USB			
Likely Questions Brainstormed And Prepared For			
ON THE DAY			
Personal Appearance			
Speak To Audience Not To Visual Aids			
Remember The 5C's For Delivering Brilliant Content			
REPETITION = RETENTION			
Last Updated			

CHAPTER 5

During the meeting

Even if it is not your meeting and you are not the meeting organiser or leader, you can still have a major influence on ensuring positive outcomes are realised.

In this chapter we suggest ways in which you can demonstrate positive qualities, through body language, presentations, group work, etc. By energising you with a new-found confidence to participate meaningfully, even meetings that you anticipate will include uncomfortable discussions and require difficult decisions will no longer be as daunting.

Right from the moment you enter the meeting room with a purposeful aura, you will judged as a valued participant, positively affecting the attitudes of other participants and the overall group dynamic.

Choosing your attitude

Consider how much time you spend doing work-related activities and, of that, how much time is devoted to meetings; not just sitting in the meeting room itself, but the preparation, the travelling, and the actions that result from them. The ability to 'choose your attitude' is therefore incredibly important as you can influence the way the meeting starts, progresses and concludes. If you enter the room believing that the meeting will be a 'waste of time' then it undoubtedly will be. If, however, you decide that your contribution and attendance will help transform the meeting into a productive, meaningful gathering – then this will motivate your fellow participants. If instead you choose to portray a different, negative attitude, you can be sure that this will have the converse effect.

'Attitude is a little thing that makes a big difference.'

Sir Winston Churchill

A 'bad attitude' is negative and can be very destructive in group situations. Before you condemn others in your own mind for demonstrating a bad attitude, make sure that your own attitude has not contributed to theirs!

Creating your positive attitude

So, what is positive attitude anyway?

You can never have absolute control over everything that happens in the meeting, but the attitude with which you choose to greet the day, approach your work, and respond to the people around you is fully within your control. Your attitude about any condition, present or future, is within your power to choose.

Your attitude is within your power to choose

brilliant anecdote

Research at the University of Texas has found that having a positive attitude to life can delay the ageing process – and that people with an upbeat view on life are less likely than pessimists to show signs of frailty.

If you think for one moment that your attitude – for whatever personal or professional reason – is not what it should be, then here is a practical way to put it right. It is so important to understand that you can consciously effect your 'moods' and that negative attitudes can be transformed – think of it as a **R A C E** that you must win.

Recognise it

Listen to the voices in your mind that tell you all is not what it should be for you. Understand that this is needlessly affecting the attitude that you are outwardly portraying.

Arrest it

Recognise the psychological mechanism that triggers these thoughts. Stop, acknowledge your position, and then make some physical move to break the negative cycle.

Change it

It could be that you stop and count to 10; better still, suggest a 'comfort break' and allow a trip to the washroom to take you physically away from the negative place you are now in. Here you can freshen up, splash some water on your face, and exit with a different perspective.

Embrace it

It is important to acknowledge to yourself that you are consciously making a choice to transform what were negatives into positives. For many people this is a very difficult change in behaviour to bring about, so if you find that you can start to do this with success, find a small reward for yourself.

It is important to practise these techniques in order to maintain your positive focus.

brilliant example

There is a very good reason why, whilst helping my five-year-old daughter to ride her first bike without stabilisers, I did not shout from behind 'mind that tree', because I knew having drawn her attention to it, that this is exactly where both she and the bike would end up – in a heap! Once a negative thought is planted it becomes the sole focus, and the tree now becomes almost the only thing she can see in front of her. Instead focus attention on a positive aspect in front; in this particular example I focused her attention on a playground in the distance. This technique can also be transferred and used to great effect in a meeting situation – instead of concentrating on a negative aspect, recognise its existence, but then refocus on a positive that you *can* control.

For the same reasons, Olympic athletes will mentally picture themselves being the 'first to cross the finish line' or to successfully complete the highest jump in a very well-rehearsed pre-event routine.

Now picture yourself with a positive attitude, participating and contributing effectively in your next meeting.

brilliant tip

According to a Stanford Research Institute study, success is 88 per cent attitude and 12 per cent education.

Respect

As part of our ACTION PLAN ground rules in Chapter 2, we suggest that you consider arriving at every meeting with the mindset that the CEO is present. If the CEO was present you would give everyone respect; you would not interrupt and talk over them, neither would you make any condescending jibes. That respect will be valued and returned to you.

However, you must participate fully and, if something is not clear, you must ask for clarification. The subject matter, the way it is being presented, and the number of people present, will determine how you go about this. Some presenters will say at the outset if they would like you to stop them in their tracks and ask if anything is unclear, others will request you wait until the end.

Notes

You might think that taking notes would be a distraction but, unless you are the official note taker, the only things that you will be noting down are those points of importance to you. So in fact, taking notes can make it easier for you to listen, and it also gives you a point of reference for later.

Questioning techniques

When you do have questions to raise, there are some great pointers to help you get precisely the information you require in a direct and succinct manner.

Closed questions

These can be answered only with a quick 'yes' or 'no', or a specific answer, and are great for obtaining facts quickly and easily as the respondent does not have to think long and hard about the answer. For example, 'Are you satisfied with your current supplier?' The control of the conversation comes straight back to you.

Open questions

Be prepared to get a long answer to an open question. The answer usually will include the feelings and opinions of the respondent, and can hand over the control of the conversation to them.

For example, 'Why has the number of damaged deliveries increased so much?'

Probing questions

These allow you to get into more detail about a specific issue or problem, and are designed to elicit opinions and rationale as well as hard facts. These are typically the 'Who, What, When, Where, How and Why' questions, which can be framed in either open or closed styles to search for more detail: 'How does this relate to what we have been talking about?' or, 'What do we already know about the way our competitors are dealing with the legislation?'

Individual or group questions

Make it clear if you are posing a question directly to an individual, the group as a whole, or for any member of the group to respond.

- For an individual, start the question with their name, 'Duncan, how will this new procedure affect your department?'
- For the whole group, maybe you want a show of hands on a proposal, so, 'Please raise your hand if you think we should talk again with that supplier.'
- If you are just looking for any response from the group, 'Who can tell me what the impact on their department will be?'

Reverse questions

Instead of answering a question, send it straight back to the questioner: 'Before I clarify my position on that, tell me about your current thinking.' This will also afford you some thinking time should you need it.

Your body language as a participant

By adopting a confident posture during the meeting, sitting upright, or on the edge of your seat, you will ensure that you remain focused on the presenter or conversation; it also indicates that you are listening.

brilliant tip

Respectfully 'mirroring' the leader's body language, that is, following the posture and type of gestures, without doing it too overtly, can make you one of the 'insiders', helping you and your suggestions get better acceptance from the group.

Avoid doodling or fidgeting. Learning to control your body movements not only means you are more focused on creating a positive overall impression, but by adopting a confident posture you will portray yourself as a confident participant.

by adopting a confident posture you will portray yourself as a confident participant

The 'arms folded' posture is often interpreted as 'defensive' or 'closed,' even when this is not your intention, so it is best avoided. Ensure that you make eye contact with the presenter; this indicates your interaction and involvement.

brilliant reminders

Participant body language

- Ensure that your posture portrays a 'confident air'. Sit up straight and on the edge of your seat whilst avoiding the 'arms folded' posture, doodling or fidgeting.
- Make eye contact with the presenter.

- Nod, and look at the speaker.

  This helps in three ways: it maintains your concentration level; it reassures those who are slightly hesitant in airing their views; and it will invariably mark you out as a 'sympathetic listener', whom others want to address.

- Remember that first impressions really do count – dress smartly.

CHAPTER 6

Blagging it!

The very nature of turning up for a meeting without being properly prepared seems to be at odds with the positive actions and preparation required for a 'Brilliant Meeting', that we are resolutely advocating throughout this book.

Whilst the ideal is for every meeting to be fully organised and thoroughly prepared for, operational circumstances and sheer pressure of work sometimes render this impossible. Faced with being unprepared for the meeting, you have a limited number of realistic choices:

- Cancel your attendance at the meeting.
- Blag it!

In some cases it may be unavoidable for you to attend a meeting that you have not prepared for adequately, due to time pressures, short notice, illness, holidays, etc., and so you will have to *blag it*. It is worth remembering at this point that amongst all the participants you are probably not alone in this situation. After all, if everyone prepared fully for every meeting they attended, there would be little time for anything else!

if everyone prepared fully for every meeting they attended, there would be little time for anything else!

We have already covered the importance of meeting preparation in Chapter 3, Before the meeting, however, should you find yourself in a meeting where you need to *blag it*, perhaps a few of these strategies may help you.

brilliant tips

Blagging it!

Bring a notebook.

Let me take the notes.

Act confidently.

Gain more thinking time by asking questions.

Gather a list of general meeting discussion topics.

Inform of your mistake.

Network during breaks.

Greatly embellish.

Is it the best option?

Take yourself out (only for the truly desperate).

Let us take a look at each of these tips in turn.

Bring a notebook with you

In this way you will always look ready to participate in a meeting, by listening actively and taking personal meeting notes and actions.

Let me take the notes

Unless they are concerned with governance, meetings today rarely have a dedicated note taker. At the start offer to take the notes, this will keep you fully occupied and deflect the need for you to provide unprepared contributions.

Act confidently

If your input is requested, first smile then give yourself thinking time with a well-practised phrase or, even better, direct a question back to the facilitator. Ask for clarification on the feedback requested, for example:

'Just to ensure that I understand what you require from me at this point, are you suggesting that . . .'

Gain more thinking time by asking questions

Make sure the questions you pose are open questions (see Chapter 5), to draw out more detailed and lengthy answers. People inherently love to talk, so whilst they are imparting their wisdom, you do not have to – *this shifts the emphasis away from you.* For those of you with children, you will recognise this well-used distraction technique!

Gather a list of general meeting discussion topics

This should be a printed list that 'gives the impression' to others that you have invested some time in preparing for the meeting. Here are some suggestions of bullet points that could appear on your list:

- Expand on ideas from the last meeting.
- Confirm related actions have been completed and review outcomes.
- Discuss timescales involved.
- Reaffirm project structure and group objectives.
- Book dates for future meetings.
- Ask for feedback from others on specific agenda items, before giving your own conclusions.
- What additional resources (external or internal) should be considered at this time?

- Do we need to involve anyone else in these meetings?

Inform of your mistake

'Sorry guys, I would rather be honest than ruin/spoil the whole meeting.'

'For some reason, the meeting was not listed in my 'Outlook' so it's a good job I ran into Dave this morning.'

Network during breaks

Use all the meeting breaks and breakout sessions to gain as much knowledge from your colleagues as possible.

Greatly embellish

Why use 10 words if you can get away with 10,000 and include some great anecdotes, trying to keep them as relevant as possible.

Is it the best option?

It is as important to know when *not to 'blag it'*, as it is to know *when to 'blag it'*. If a direct question has been posed in your direction, and no amount of thinking time will give you the answer – then the best advice is to be honest. Repeat the question (to confirm you have understood it correctly) then propose that you take this away as an action, adding that you will collate all the relevant information to present at a later date, or the next meeting.

Take yourself out (only for the truly desperate)

Organise for a colleague to interrupt the meeting that you are attending, with an urgent phone call/task that will take you out of the meeting and away from your 'unprepared' situation.

Conclusion

Being unprepared is not a comfortable position to find yourself in. However, as a consequence of today's increasingly busy and long working days, it is somewhat inevitable that you will, at some point, be less prepared than you would like, as you enter a meeting environment. We hope that these ideas will allow you to make the best of every meeting that you actively participate in.

CHAPTER 7

After the meeting

Once the meeting is over, the actions and initiatives agreed upon by the participants can now begin; in effect, the 'real work starts here'.

This assumes of course that you have not just experienced another of those 'meetings about meetings' where nothing advances. In fact, if this is the case, then the meeting should not have taken place at all.

Leaving the room

In a busy organisation, chances are that no sooner has your meeting finished then the next one in that room is going to begin. In larger organisations there may be someone assigned to clear everything away and set the room out for the next meeting. Nonetheless, it is no more than a common courtesy to remove from the table your own debris; the used glass or cup, any waste paper, etc. Are there any meeting notes still on a flip chart or dry erase board? Even if they are not confidential or would mean nothing to others, by leaving them visible you are inhibiting the immediate progress of that meeting. Remove the paper from the flip chart or turn it over to leave a clean page showing; erase the ink from the dry erase board. The simple rule is to leave the room in the exact state that

> leave the room in the exact state that you would have expected to find it

you would have expected to find it at the start of your meeting. And if everyone does a little to help, it really will not take long.

Another common practice when the meeting has concluded is for participants to hang around in the room talking in pairs or small groups. Maybe it is last night's football or legitimate business, either way we recognise that now the meeting has finished people are going to be departing in different directions, and so we take the opportunity to have these catch ups. Again, it is the next meeting you potentially are stopping from getting under way, so why not exit the room and find somewhere informal to continue those time outs?

Meeting actions

If this is a more formal type of meeting, you should receive the notes soon after the meeting has ended, which clearly set out individual actions, expectations and timescales. However, if you do not have this luxury, ensure that you have noted your personal actions clearly, and any other actions assigned to group members that could impact on the effectiveness and timescales of your results.

Notes will also refresh your memory regarding the presentations and discussions that took place, which can be essential if you have more than one meeting per business day.

The actions assigned to you are part of your job, so first assess how long each action will take you to complete. If the next action can be done in two minutes or less, do it when you first pick up the item.

Schedule the rest of your actions in to your calendar now so that

 tip

David Allen's two-minute rule

If the next action can be done in two minutes or less, do it when you first pick up the item – even if that item is not a 'high priority' – because it takes longer to store and track any item than to deal with it the first time it is in your head. See www.gtdtimes.com.

you are regularly reminded. What will the impact be on others if you are late completing your actions? And by the same token what actions do group members need to complete before you can commence with yours? It is very easy to look at our work in isolation without realising the impact it has above, below and alongside us. If you have any doubts or concerns, then communicate this well in advance. If you have other work to do, then ask what will be the consequence of completing your task later than scheduled.

To make sure that you will not need to read the information in Chapter 6, 'Blagging it', schedule the dates for the next meeting (if there is one) in to your calendar.

Communicating outcomes

If you are a designated representative for your colleagues or group, you need to be very clear about how you are going to share with them the outcomes from the meeting. Is it not human nature that we all assume that no news is bad news? So even if the news is not as positive as everyone is expecting, it is still better to share it with them and put a positive spin on it. Not communicating formally is how 'Chinese

Not communicating formally is how 'Chinese Whispers' start

Whispers' start, and they are sure to spread more quickly in the absence of any formal information!

Depending on the number of people you need to communicate with and where they are located, will determine how you best share the news with them. An email is going to be the quickest way and, with today's communication, it is likely that the message will be transmitted to their mobile phone screen even if they are not at their desks. However, if a very important message needs communicating, then a personal message delivered by you is best to alleviate any misinterpretations or concerns that may result from a quick email.

Can you group everyone together for a quick conference call? If people are in different time zones why not record them a personal audio (MP3) or video message they can play on their desktops or mobile devices as soon as they are back at work? This technology is readily available; it is inexpensive and will not overload IT networks, and it is what we have come to expect in the YouTube™ generation.

Summary for participating in a Brilliant Meeting

Review the agenda	Propose any additional items.
Understand the meeting purpose	Ensure that the objectives and proposed outcomes are clear and identify how you can personally contribute to achieving these.
Elevator Pitch	Develop your own personal, compelling Elevator Pitch that is *CONCISE*, *PROFESSIONAL* and *RELEVANT*. Practise this until it becomes natural.
Plan and prepare	Content/contributions/presentations as appropriate whilst reviewing your personal action items from previous meetings.
Content	When preparing any content to be used in a meeting remember these simple ABCs:
Appealing	In terms of content and delivery.
Believable	Everything should be factual and deliverable.
Clear	Use positive language free from jargon.
ACTION PLAN	For use during every Brilliant Meeting.
Arrive	On time with relevant, well prepared content.
Choose	Your attitude.
Turn off	All personal communication devices; phone, Blackberry™, laptop, etc.
Imagine	Your CEO/MD is present.
Obey	The agenda and stay until the end of the meeting.
Never	Use jargon and avoid distracting side conversations.
Participate	Actively.
Learn	What you do not know, share what you do know.
Accept	And fully support consensus decisions.
Named	Actions must be completed.

Use the checklist to ensure your successful participation in future meetings. Download it from www.meetingexpert.co.uk.

Checklist for Participating in a Brilliant Meeting

space2inspire

Summary Information
Meeting Title
Name of Group
Date & Time of Meeting
Meeting Venue

	Done	Comments	Date / Ref
Review Agenda			
Understand Meeting Purpose and Objectives			
Accept / Decline / Delegate Meeting Invite			
Suggest Additional Agenda Items			
Personal Introduction – Elevator Pitch			
Plan Contribution			
Complete Pre-Meeting Assignments			
Ensure Previous Meeting Actions are Completed			
Plan to Arrive Early			
Choose Attitude			
Meeting Notes			
Evaluate and Give Meeting Feedback			
Communicating Meeting Outcomes			
Begin Work on Assigned Action Items			
Other			
Last Updated			

PART 2

Next time you prepare for a Brilliant Meeting

Introduction

Before you begin to prepare for any meeting, ask yourself this fundamental question:

'Is a physical meeting the most effective method for achieving the desired outcomes?'

If the answer is 'No' then do not organise a physical meeting – consider your alternatives.

If the answer is 'Yes' then the foundation for a Brilliant Meeting begins with preparation.

brilliant tip

Time spent in meetings *will* be reduced through effective preparation.

The skills required to organise a Brilliant Meeting are quite different to those needed for leading one. If you have 'called' the meeting, the probability is that you now have the combined responsibilities of leader and organiser. Even if the latter is with assistance from a secretary or PA, there is a real danger that, if essential preparation is not adequately completed, the result will be a 'poor meeting' for all involved.

Meetings are vital to organisations and, in order to facilitate participants working together as a group, much work will be required to ensure that the 'Why, Who, When, Where, What' for the meeting have all been completed prior to the event, whatever the subject matter, meeting type or purpose.

So whether it is a project meeting, group meeting or a brainstorming session, this section contains suggestions to assist with preparation and help transform your next meeting into a Brilliant Meeting!

quote

'By failing to prepare, you are preparing to fail.'

Benjamin Franklin

CHAPTER 8

Why has the meeting been scheduled?

Meetings are just one of several tools of communication used within and between organisations. Used correctly they can be the most important conduit for effective and productive knowledge exchange. In a structured environment with the right people purposefully involved, meetings are where we all have the opportunity to learn, share and create.

meetings are where we all have the opportunity to learn, share and create

Meeting purpose

It is important to identify and detail the purpose for holding the meeting to all involved. The correct environment for the meeting can now be selected, and the participants can focus their preparations accordingly.

The term 'Meeting' has come to be used as a catch-all for many different situations when groups of people get together. These include, but are not limited, to:

- information dissemination;
- discussions, leading to conclusions/achieving objectives;
- brainstorming;
- training/workshops;
- planning;

- giving feedback/performance reviews;
- process mapping;
- product development;
- knowledge exchange;
- dispute resolution;
- contract negotiation;
- change management;
- obtaining opinions and feedback.

All of the meeting purposes listed fit in to one of three categories, which in turn will help set the meeting objectives.

tip

Ask yourself if the meeting is being scheduled to enable *learning*, *sharing* or *creativity*?

Three meeting purposes

- Learn – presentation, information dissemination, organisational meeting, etc.
- Share – status updates, financial reviews, sales meetings, discussions, etc.
- Create – problem solving, process mapping, product design, brainstorming, etc.

Even with both the style of meeting and also a seemingly clear purpose identified, if you are still unsure about the absolute necessity of a physical meeting (same time, same place) – then ask yourself a few searching questions to avoid the trap of 'a meeting for meeting's sake':

Could an alternative be more effective?

Could the meeting purpose be supported as effectively by using an alternative to a physical 'same time, same place' meeting?

- Emails bombard our 'Inboxes', but a well-written concise attachment could work just as well.
- Blogs and wikis are becoming effective tools for groups to formally share information at any time, without the formality of a meeting, although these methods of data transfer are not effective if real-time collaboration is necessary.

brilliant definition

A *blog* (a contraction of the term *web log*) is a Web site, usually maintained by an individual, with regular entries of commentary, descriptions of events, or other material such as graphics or video.

A *wiki* is a collection of Web pages designed to enable anyone who accesses it, to contribute or modify content at any time.

- Webcasts published on the organisation's intranet deliver information that can be viewed on demand.
- Podcasts published on the organisation's intranet can be downloaded and listened to when and where suits the individual best.
- Internet/intranet connections using tools such as NetMeeting™, SameTime™ and WebEx™ allow the sharing of data between two or more connected computers.

brilliant definition

NetMeeting™, SameTime™ and WebEx™ are software applications from different manufacturers that allow users to collaborate over distance sharing text, images and, in some cases, control of the application via the public Internet.

If you are making content available via any of these methods, be certain that everyone can access it easily and effectively. For the younger generation entering the workplace, digital information is a given, for others the ability to access the information may not be so straightforward.

brilliant tip

Quite simply, if there are viable alternatives, do not meet when:

- participants do not need to interact in person;
- participants do not need to have the information at the same time;
- communication does not need to be two-way in real time;
- you do not need to see personal reactions conveyed through body language.

What if the meeting did not happen?

- Would any part of the project be jeopardised if a physical meeting did not take place?
- How would participants react if the meeting did not go ahead?

Be aware that, if you start using the telephone, email or Web to replace meetings, then having too few meetings can be as disruptive as having too many, but an alternative to a physical meeting should always be considered.

Meeting objectives

A successful meeting is our ultimate objective but in order to gauge effectiveness we need a set of objectives against which we can measure the accomplishments. Furthermore, clear objectives focus the group on achieving those outcomes within the time period scheduled for the meeting, so long as the set objectives are 'SMART'.

brilliant tip

Use this simple SMART acronym to set worthwhile and meaningful objectives.

1. **S**pecific – Objectives should clearly specify what you want to achieve.
2. **M**easurable – You should be able to measure whether you are meeting the objectives or not.
3. **A**chievable – Are the objectives you set achievable and attainable?
4. **R**ealistic – Can you realistically achieve the objectives with the resources you have?
5. **T**ime – In what timescale do you want to achieve the set objectives?

Meeting title

Communicate the meeting purpose – clearly and concisely – in the Meeting Title, so participants have the relevant information to decide whether their acceptance of the invitation is the right course of action. If the meeting is of a 'confidential' nature, consider the wording of the meeting title carefully, as information could be displayed in electronic schedules, etc.

The agenda

The word agenda literally means 'things to be done', and one of the most important aspects of a successful meeting is a carefully constructed

one of the most important aspects of a successful meeting is a carefully constructed agenda

agenda. This document summarises the structure of the meeting whilst also detailing the purpose, objectives and desired outcomes in a given timescale.

There will be occasions when legal requirements and procedures govern the agenda of a meeting as well as some of the agenda items, such as election of officers, appointment of accountants, etc., but for the purpose of this section we are going to assume that our Brilliant Meeting is not subject to such rules.

If you have the luxury of sufficient time, then circulate a note to proposed participants with an outline of what the meeting items will be, and ask now for any feedback and proposed additions. The reality for most meetings is that a meeting invitation is issued long before the agenda has been finalised, and it is only shortly before the meeting itself that the agenda is circulated, thus not allowing for any additional items to be submitted from participants.

Essential meeting logistics

These items are individually covered in more detail later in this section; however the following important logistical arrangements need to be detailed in the agenda documentation:

- the date, start and finish times;
- venue;
- meeting leader;
- invited participants;
- the dress code for the meeting, if it is different from the norm.

brilliant tip

For external participants or those located at other offices, do you need to include directions and transport alternatives? Perhaps the meeting venue has a Web page that you can link to?

Constructing the agenda

Sequence

Items that need fewer than 10 minutes to complete along with the 'less important issues' should be placed at the very top of the agenda, but you must be sure that these can be concluded successfully within the time constraints. This will result in those items being dealt with quickly, before the meeting moves on to the larger, more important items, and avoids the smaller items 'falling off the bottom' of the agenda.

Urgent items should be prioritised towards the beginning of the meeting. However, try not to schedule 'heavy' and controversial subject items close together. This will encourage a diversity of agenda items and help with the flow and intensity of the meeting.

Item allocation

Each agenda item should identify who is leading the discussion/presentation, along with the desired outcome. If necessary prepare a document with fuller descriptions and more background information, or consider circulating a separate briefing note with the agenda.

Timings for individual agenda items

Clearly set out the order in which items are to be discussed, together with the time allocated for completing each item.

Breaks and catering

Define when and for how long breaks have been scheduled, along with information about the refreshments that will be provided.

Meeting duration

In order to achieve the purpose and objectives that you have already set out, be realistic when determining how long the meeting should be scheduled for. Remember to take comfort and refreshment breaks into account.

brilliant tip

Schedule the time realistically needed to complete the agenda, as opposed to making the agenda fit to the time.

Two agendas

Consider developing two agendas for the meeting: one is distributed to invitees, and the other is for the meeting leader to help keep the meeting on track. The public agenda lists topics, speakers and allocated time. The leader's version ranks the agenda items in order of importance (not necessarily in running order) to highlight those items that *must* be covered before the meeting concludes.

Format

Leave blank spaces below each line of the agenda. This allows participants to make their own notes on the agenda document itself, allowing even the most unprepared participant the opportunity to record pertinent information from the meeting for themselves.

Date of next meeting

In order to save time organising future meetings, ensure this agenda item is listed – particularly on the leader's copy of the agenda as a 'must do'. This will also encourage the participants to consider their own availability when committing their attendance to a future meeting.

Issuing the agenda

The most common way to send the agenda to the proposed participants is by email.

brilliant tip

As an alternative, consider posting an electronic version of the agenda on to an intranet page and email those people with secure access a link to that page. This will ensure that, as the agenda develops, participants will always have access to the latest version without a constant barrage of emails.

The 'Any Other Business' (AOB) trap

Do not list AOB as an agenda heading – this will just elongate the meeting with subjects that have not been prepared for. Instead ask for AOB contributions in advance, so that time can be allocated on the main agenda.

Achieving outcomes

A successful meeting has occurred when all the agenda items and agreed outcomes have been reached, with clear actions, responsibilities and deadlines – all completed in the time allocated!

brilliant example

Space 2 Inspire Limited

Strategy Meeting

Date: Wednesday 13 May 2009
Time: 11.00–12.55 followed by lunch
Venue: Central Boulevard, Oxford – www.space2inspire.co.uk/locations
Meeting Leader: Duncan Peberdy
Participants: All Board Members
Dress Code: Business casual

AGENDA

11.00–11.10 Coffee

11.10–11.20 Welcome from Meeting Leader
Apologies, ground rules, notes from last meeting

11.20–11.40 2008 Business review – JH

11.40–12.15 Discussion of Top 6 Accounts by revenue – All
See attached spreadsheet on revenues by account

12.15–12.40 Strategy for new busines – DP/JH
Key Account Management, credit crunch implications

12.40–12.55 New business pipeline, marketing and opportunities – GD/JH
GD to present new marketing campaign
Set date of next meeting

13.00– Lunch then depart

Documents attached Draft Management Accounts to 31 March 2009
Newspaper article on business meetings

Costs and benefits

Is it possible to calculate an approximate cost of your meeting?

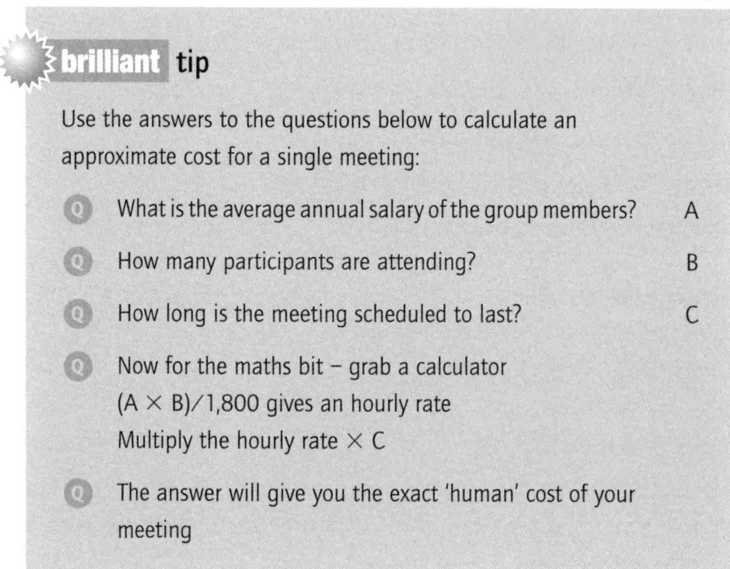

brilliant tip

Use the answers to the questions below to calculate an approximate cost for a single meeting:

- What is the average annual salary of the group members? A
- How many participants are attending? B
- How long is the meeting scheduled to last? C
- Now for the maths bit – grab a calculator
  (A × B)/1,800 gives an hourly rate
  Multiply the hourly rate × C
- The answer will give you the exact 'human' cost of your meeting

The temptation might be to believe that because everyone is 'on hand' there are no significant costs associated with your meeting but, as illustrated in the following example, costs can be significant.

brilliant example

Average annual salary of group members	£35,000	A
Number of participants	6	A
Scheduled meeting duration	2 hours	A
The 2-hour meeting *will* cost at least	£234.00	A

£35,000 × 6 = £210,000 Annual salary of group

£210,000/1,800 = £117 Hourly rate

£117 × 2 = £234	Cost in 'man hours' + employee benefits + employer tax/NI, etc.

Now consider the hidden, but very real infrastructure costs of the following:

- the meeting room itself;
- the furniture and equipment in the room;
- catering and refreshments.

Intangible benefits

Of course there are some benefits that you cannot put a cost against. These are listed below.

Team building

To really feel part of group you do occasionally have to meet up. People achieve more when they work together as a group with a common goal.

brilliant quote

'Coming together is a beginning
Keeping together is progress
Working together is success.'

Henry Ford

Motivation

If you are fortunate to have a leader who is motivational just by their words and presence, then having them in the room will deliver far more than sending out their message in an email. If they have a commanding voice this will come across on a conference call, but nothing will surpass being there with them in person.

On a personal note

> People achieve more when they work together as a group with a common goal

Time during breaks, before and after meetings, can provide an opportunity to 'get to know' your colleagues on a more personal level. This will benefit the professional interactions of the individuals through better personal understandings and relationships.

On their own these invisible benefits are not sufficient enough to determine that a 'same time – same place' meeting is a must, but they should be part of the decision-making process when plans are being made to hold a physical meeting.

CHAPTER 9

Who needs to be invited?

Choosing your participants

The meeting objectives have been determined and, in order to accomplish them, input is required from numerous contributors. The right balance of participants needed to achieve the meeting goals must now be decided upon. When selecting the participants, consider job roles, experience and perhaps even personalities, all with the end goal in mind. With too few people involved, a decision may be reached quickly, but was the breadth of knowledge, experience and stakeholder involvement present, to ensure support and implementation? With too many participants the time needed to hear and consider input from everyone might prove too long. Within a larger group, there is an increased likelihood of smaller fragmented groups forming, leading to difficulties in reaching consensus decisions and maintaining control of the group.

The purpose of the meeting will determine the participants whose presence is necessary in order to achieve the objectives and these will include the following:

The purpose of the meeting will determine the participants whose presence is necessary

- subject matter experts;
- ultimate decision maker(s);
- those responsible for implementation of any decision;
- direct stakeholders and those affected;

- HR or employee representatives where any arbitration or conciliation is required;
- experience – perhaps even in a non-executive capacity.

Except for those who have the luxury of optional attendance – such as executive sponsors – the participants selected to attend are those who have a direct role in the proceedings and a stakeholding in the outcome.

Overcoming physical barriers

Just because a chosen participant is based remotely should not preclude them from being invited to the meeting. The preferred method of inclusion would be to use video conferencing, allowing the remote participant to take an active role in the meeting, whilst being visible to all. As well as video, there are other conferencing options through which remote participants can be included effectively, such as telephone and data conferencing. Chapter 13 provides an overview of available equipment and guidance of best practices.

Attending only part of the meeting

Consider inviting subject matter experts, internal and/or external, for only the section of the meeting that is relevant and valuable to both them as individuals and the group. Try to schedule these participants close to a break in the meeting to limit potential disruptions.

Personalities

Managing participation once the participants are assembled is the job of the meeting leader, and in Chapter 15 we take a closer look at the differing types of personalities that could be sitting in the room.

An ideal group dynamic will be a blend of personality types. A meeting room full of extroverts may be difficult to manage as everyone competes to have their opinions heard and, by the same measure, a room full of introverts might challenge even the best leader to encourage meaningful participation.

> An ideal group dynamic will be a blend of personality types

Who to leave out

Coinciding with the introduction of Brilliant Meetings, now is the ideal time to have a hard look at who has been attending meetings, and who should be attending meetings in the future. Just because someone has always taken part in a certain series of meetings, does not mean that they should automatically be included in the future.

brilliant tip

Perform a 'meeting audit' to evaluate the contributions made by each participant. To download a meeting audit checklist visit www.meetingexpert.co.uk.

In order to assist in the decision-making process of who should be included in a forthcoming meeting, review the following list.

- From the previous meeting notes, review contributions made by individual participants. Does their name appear regularly in the meeting notes, either for contributions made during the meeting or because they have been assigned actions?
- Consider what preparation these people make before a meeting. Are they motivated to participate actively and carry out the required pre-meeting preparation?
- Rate their overall value to the meeting.
- What value do they take away from the meeting?

- Where and with whom do they share the outcomes of the meetings?
- What would be the impact on the group dynamics if an individual was not invited to the next meeting?

If it is decided that the presence of an individual is no longer required, suggest that, instead of inviting them to future meetings, it would be a better use of their time to receive a copy of the meeting notes only.

Assign meeting roles

For some meetings, tasks need to be delegated before the meeting starts, so that participants arrive knowing what additional role is expected of them. Whether it is a task to ensure that the meeting leader is keeping to time, or the recording of questions and answers given, all these additional responsibilities should be decided upon before the meeting commences. This will also help to define any additional equipment needed, such as flip chart, whiteboard, projector, etc. (See Chapter 13 for additional information.) Not all of the following roles will be appropriate for every meeting, in fact for many less formal meetings it may be only a meeting leader that is required, but they should be considered as follows:

- meeting leader;
- note taker;
- scribe to write notes/record questions on flip chart or interactive whiteboard;
- timekeeper;
- independent facilitator;
- presenters;
- audio visual technician.

If these or any other roles are to be assigned to participants, get their approval beforehand – a quick phone call or email should suffice – so that the meeting invitation and other information can be circulated with details of these duties included.

CHAPTER 10

When should the meeting be scheduled?

Meeting date

Determining when to hold the meeting needs careful consideration. Heavy work schedules make it increasingly challenging to find a time that suits everyone perfectly and, the more senior the participants are, the less available they tend to be.

the more senior the participants are, the less available they tend to be

Of course there are many factors that will have an impact on the meeting date. These might include the following:

- the venue and its availability;
- where the participants are located and their travel considerations;
- how quickly the briefing material can be gathered and circulated together with the agenda.

Dates for regular project, sales and board meetings are typically agreed and booked up several months in advance – this reduces costs by allowing participants enough notice to take advantage of offers associated with the advance booking of travel and accommodation. If the majority of the group has to stay overnight it is often worth encouraging the remainder to do so as well, since the 'team building' benefits associated with an evening socialising together are considerable.

Overnight accommodation for all, on the evening prior to the meeting, also allows for a much earlier and prompt start.

For one-off meetings that are called at short notice, getting the time and venue to suit everyone is more problematic.

If time allows, offer alternative dates to the key participants and build up the meeting from there.

These are all important elements but, above all else, you need to be mindful of the timescale required for the actions and results to be reached. A crisis meeting called at short notice will not have the luxury of advance preparation, but let us assume that we have time to organise a Brilliant Meeting.

Travel

Take into consideration who will be invited and how far they have to travel. If participants have less than a two-hour journey, then it is unlikely that they will travel the day before. If the meeting is scheduled to commence at 9 a.m., then chances are they will be leaving home shortly after 6 a.m. – to build in buffer time for rush-hour traffic and unforeseen circumstances. Will you get the best out of these participants in this circumstance, or is this just an unavoidable facet of modern business life?

If the meeting requires a specific environment, or the meeting includes an external visit, then the time of the meeting could be governed by these outside influences.

Time of day

Certainly a morning meeting is generally considered to be preferable for increased alertness and productivity, and is desirable if the meeting is scheduled to last for five hours or more. However, as well as a published start time, Brilliant Meetings should have a published end time. An 'on time' finish is more likely to be achieved if the meeting is scheduled to conclude just before lunch or close of business.

brilliant tip

Publish meeting start and end times – and keep to both of them!

Avoid booking a short meeting in the middle of the day as this can waste a lot of time in travel, therefore reducing the productivity of the participants in their other duties. Breakfast meetings are a good idea in certain cultures, but could be considered too demanding on private schedules in some cultures. Do not forget also to consider people's travelling times after the meeting; if a late finish time is scheduled, think about offering overnight accommodation if warranted.

brilliant tip

Time limits create a sense of urgency and focus that can greatly aid meeting productivity.

Here is another idea on scheduling the time for a meeting.

brilliant tip

Publish a memorable start and end time for your meeting. For example:
Thursday 27 August, Asset Management Meeting, Boardroom
Start 11.11 a.m.
Finish 12.26 p.m.

Most meetings start on the hour or half-hour and we have become totally conditioned to this. Now think about starting and ending the meeting at curious times. So instead of an 11 a.m. start choose 11.11, with a finish of 12.26. The most recent versions of Microsoft Outlook™ allow specific times to be entered in to the meeting invitation and calendar.

Duration

There are too many issues that can affect the time required for an effective meeting and there is no magic formula to calculate the correct meeting duration – that we know about!

A great starting point for calculating meeting duration is first to develop a draft agenda and to allocate a notional time to each item. This may seem obvious but how many meetings have you attended where the meeting has over-run and/or the agenda was not completed? Estimating start and finish times for each item will give you confidence that the agenda can be realistically completed within the allocated time. This important process should not take an undue amount of time, but is worthy of your full consideration. (Chapter 8 gives guidance on constructing an agenda.)

Breaks

Scheduled breaks should be included in the calculation for the duration of the meeting; the longer the meeting – the more breaks necessary. If people are not participating fully all of the time, concentration begins to drop after just 45 minutes. 'Comfort breaks' also need to be scheduled regularly, so plan for at least a 'five-minute break' every hour. For prolonged meetings, consider also the need to allow participants time to access phone messages/emails, etc.

Presenters

If the meeting incorporates external presenters who will participate in the meeting only for the duration of their own presentation, then consider carefully at what point they are scheduled. If set up of computer or other equipment is required, there is a danger that participants will view that time as downtime leaving them with nothing to do. Once the meeting continuity is broken, participants may use the opportunity to visit the toilet, start side conversations, check emails, etc. Consider scheduling the presenter to start immediately after a timetabled break, so that equipment can be configured whilst the meeting participants are out of the room.

Set up time, clear up time

If the meeting space is reserved using a computerised diary system, then the start and end times are broadcast automatically to the participants. If you need to set aside time before and/or after the meeting to make any adjustments to the room layout, or to have equipment prepared, you will need to book additional meeting times as appropriate. Otherwise, participants will be getting in your way for the first 10 minutes, which will be a complete waste of their time too.

Time

There is a tendency to book a meeting space for longer than required, 'just in case'. If the agenda determines that 90 minutes is the right length of time to complete it, then book the space for 90 minutes. If instead the space is booked for 2 hours, then doubtlessly the additional 30 minutes will be filled. However, the meeting will

> time constraints help to focus solely on the topics in a concise, productive way

suffer as a result; time constraints help to focus solely on the topics in a concise, productive way. Instead, people will go off topic, and trivial peripheral discussions may creep in.

CHAPTER 11

Where should the meeting take place?

The choice of venue is more important than many people believe.

The choice of venue is more important than many people believe

'Won't any old room do?' Absolutely not.

Although many meetings are relatively informal and held 'on-site' in dedicated meeting rooms, the meeting type, numbers of participants and room configuration are all factors that have a huge bearing on the room choice, and therefore form part of the overall effectiveness of the meeting.

Venue considerations

Critical to the success of a meeting is matching the venue to the meeting purpose, taking into consideration the following:

- the size of room for the chosen number of participants;
- the room configuration, type and comfort of furniture;
- audio visual equipment and other in-room resources;
- heating, lighting, ventilation requirements;
- potential for disruption/distraction;
- need for confidentiality/neutrality;
- disabled access;
- travel/environmental factors;

- accommodation/hospitality;
- availability and cost.

Standard meeting rooms typically are configured in traditional boardroom style. An oblong table with chairs placed around the outside, possibly equipped with a projector, screen and flip chart easel. For some meeting types, this traditionally configured space will reduce the chances of successfully achieving the meeting objectives and purpose.

Because of meeting type and purpose, the following examples require additional venue considerations;

- **Collaboration/Brainstorming**

  If succinct, productive collaboration is the desired outcome, then the room must be the correct size for the number of participants. The meeting room should be configured to allow equal access by all participants to any group device, such as a flip chart, whiteboard, etc. Collaboration does not work well in a theatre-style environment. People need to have line of sight to each other's eyes, they need to be close enough to feel included, but not too crammed in that they are invading the next person's personal space.

  If collaboration over distance is required, then the venue choice will be subject to correct data connectivity to allow all locations to see, share and participate effectively.

example

Collabatorium

Currently a UK university is planning to develop a 'Collabatorium' for brainstorming meetings – a dedicated environment free from interruption and technology problems.

The principle is that participants have to be available for the entire meeting duration, and then they are metaphorically locked into the room until the outcomes are reached. Video and data conferencing equipment allows other locations and people to be connected for any required distance collaboration.

The environment will be complete with a fridge, into which lunch or other refreshments (if required) will be placed before the meeting starts, a coffee machine supplies vital caffeine, and the room has it own toilets leading directly from it.

This experience limits the likelihood of interruption and external distractions, allowing the participants to focus solely on the task in hand.

- **Confidential**
  If the content of the meeting is highly confidential then do not book an open environment or a room with glass walls. Also remember to schedule time to remove confidential information from whiteboards and flip charts when the meeting has concluded.
- **Sharing information**
  If formal presentations are to take place as part of a meeting designed to share information, then the room should be set up in a theatre-style configuration, with the audience in rows, preferably with tables, facing the meeting leader.
- **Training/workshops**
  Traditionally a training environment is set up 'classroom style' with rows of tables and chairs directed towards the front of the room. Information is broadcast, and notable points captured separately on flip charts or dry erase boards. However, rooms now exist that allow multiple screens of information to be displayed simultaneously, encouraging the content to be moved, annotated and captured in a digital format and in real time.

brilliant example

Microsoft's PowerPoint™ is traditionally the software used to deliver training sessions, due to familiarity and ease of use. However, the content is then constrained to a 'single window' or one slide of information at a time, reducing the persistence and significance of the material.

Multi-Slides is a solution that allows PowerPoint™ slides to be simultaneously arrayed around the room that delivers multiple benefits of information persistence, digital annotations and, most significantly, ease of use in any training/meeting environment.

For more information visit www.meetingexpert.co.uk

- **Informal**
  To create a relaxed atmosphere, conducive to planning and creativity sessions, organise lounge-style furniture, with easy chairs and coffee tables.

If the only space available within your organisation is not right, then either wait until a suitable space becomes available, or hold the meeting off-site at a hotel or conference location. Yes this will incur an additional cost, but the cost of a poor meeting in your own offices will be greater still.

brilliant tip

Always go and check the room personally, in advance.

Do not wait until the day itself only to find that it does not suit the needs of the meeting.

'More work, less traffic'

Connected meetings held simultaneously in multiple venues can be delivered successfully through modern conferencing solutions. Accessing technologies such as video conferencing may dictate the venue for the meeting and therefore may also have an impact on your meeting budget if you need to book external meeting facilities. However, video conferencing consistently delivers benefits such as:

- increased productivity through quicker engagement;
- elimination of expensive and extensive travel;
- reduction in carbon footprint, in line with corporate social responsibility targets;
- improved work/life balance.

Facilities

Dependent on the meeting format, the following services and resources need to be considered and, if required, confirmed at the same time as booking the venue:

- catering;
- table and seating type and configuration;
- projector and screen;
- video conferencing;
- audio conferencing;
- flip charts and marker pens;
- dry erase board and pens;
- interactive whiteboard and relevant software;
- access to in-room computer;
- microphones;
- response systems;

- visualiser;
- VCR/DVD.

If your meeting room needs setting up in a specific arrangement, ensure that the set up time is included in your room booking.

For information on how to get the best from your existing in-room resources see Chapter 13.

CHAPTER 12

What else needs completing?

Reinforce ground rules

> ground rules must be clear, consistent, agreed-to and embraced

A Brilliant Meeting includes ground rules that are developed and used by groups, but preferably adopted throughout the organisation. These rules are statements of values and guidelines that a group consciously establishes to help individual members decide how to act during a meeting. To be effective, ground rules must be clear, consistent, agreed-to and embraced by all participants.

example

Brilliant Meeting ground rules

Arrive on time, with relevant, well-prepared content

Choose your attitude

Turn off all personal communication devices: phone, Blackberry™, laptop, etc.

Imagine your Chief Executive Officer (CEO) or Managing Director (MD) is present

Obey the agenda and stay until the end of the meeting

Never use jargon and avoid distracting side conversations

▶

Participate actively – 'silence is acceptance'

Learn what you do not know, share what you do know

Accept and fully support consensus decisions

Named actions must be completed

See www.meetingexpert.co.uk for more suggestions and examples.

It is worthwhile detailing these meeting ground rules in the meeting invitation, either as a link to the organisation's intranet (if available), or as a document that can be printed (for external participants), to reinforce their inclusion as part of the meeting process. The meeting ground rules should also be displayed at the meeting venue to remind participants what behaviour is acceptable and that these rules are positively encouraged as part of the values of the organisation.

Distribute briefing notes

Ensure that any information that needs to be accessed before the meeting is sent to the participants early enough to allow sufficient time to complete any necessary reading or further research. Be careful of issuing, in advance, a full transcript of the information as the consequences could include: a reduction in the effectiveness of the presentation or discussion, a reduction in the impact and, ultimately, a reduction in participation. Briefing documents include, but are not limited to: reports, drawings, proposals, technical certifications, financial statements and projections.

Briefing information may need to be collated from a variety of sources. However, in order to keep the material relevant and concise, it is worth detailing the precise format required for inclusion. For example, you could insist that only the facts are presented. Alternatively you could allow the briefing papers to present the facts and propose a number of scenarios on the way forward. However, be aware that this type of persuasion might constrict an open discussion and hamper further idea development.

What is important, for non-subject matter experts, is that the briefing papers:

- are concise;
- state the key message clearly;
- act as an executive summary rather than in-depth specialist detail;
- are clearly presented for maximum impact.

brilliant tip

Invest in a copy of Jon Moon's book *How To Make An Impact.* This is essential reading for anyone wanting to influence, inform and impress with reports, presentations and business documents. www.jmoon.co.uk

Whilst we do not suggest the trivialisation of strategically important issues, briefing documents must be suitably brief and accessible to non-specialists or they risk not being read at all.

Collate presentations in advance

brilliant tip

To ensure that all presentations run smoothly during the meeting, especially if they are to be combined into one, send information on file formats and timescales by which any material should be submitted by contributors. For example:

PowerPoint™ presentation in version 2007 or earlier

No later than Monday 1 December, emailed to ....

brilliant tips

Presenters all too readily fill their PowerPoint™ slides with images, thereby making the file sizes too large for sending via email, and slowing transitions when delivering from a USB memory stick.

If you are collating all the PowerPoint™ presentations in advance of the meeting, ensure that the authors reduce the file sizes before emailing them to you.

To compress images within a PowerPoint™ presentation (which generally does not reduce the visible quality) do the following, which will significantly reduce the overall file size:

On any PowerPoint™ slide select an image by clicking with the left mouse button. Now click with the right mouse button and choose 'Properties' and 'Format Picture'. Choose a lower output resolution suitable for Web pages and projectors, and select the option to apply this change to all images in your presentation.

Now when you re-save the presentation, the file size will be greatly reduced; just perfect if presentations are being sent by email.

Collating presentations in advance of the meeting will ensure smooth presenter transitions and minimal meeting disruption; after all it is suggested that it takes 10 minutes, on average, for a meeting to 'get back on track' after an unscheduled break or interruption.

Schedule catering and breaks

Refreshments

Refreshments provide participants with the necessary energy to maintain their attention; typically this manifests itself as coffee and chocolate biscuits. As much as these are anticipated, they are not necessarily the best choice for the meeting or a great choice from a nutritional point of view. Your participants may not immediately embrace the decision to replace the chocolate biscuits with light healthy snacks, but in fact replacing high-sugar foods with snacks such as fresh fruit and cereal bars will result in a prolonged energy injection, rather than the high peak and low trough of sugary foods.

Refreshments provide participants with the necessary energy to maintain their attention

brilliant tip

To maintain energy levels, replace biscuits or pastries, which are energy-dense and calorific, with foods rich in natural sugars, such as fruit and cereal bars, which are absorbed slowly, without a big sugar rush.

Scheduling refreshments can also be used to gently encourage participants to arrive at the meeting space on time – complimentary fresh coffee should be more enticing than having to purchase a vending machine alternative. This may also stop participants from wandering back to their desks during natural breaks and lunch.

Lunch

If lunch is required, first check with the participants for any special dietary requirements. If possible, schedule the food to be delivered to a location outside of the meeting room, thus minimising disruption to the meeting. Ensure that timings are detailed to both the caterers and also the meeting leader, especially if hot food is being provided. Also schedule when the lunch break is to end, so that the catering staff can ensure the leftover food and crockery are removed in good time.

Breaks

Concentration begins to suffer after 45 minutes of uninterrupted sitting

Concentration begins to suffer after 45 minutes of uninterrupted sitting. Schedule breaks where participants are encouraged to get up and move around for at least five minutes.

Scheduling breakout sessions

Breakout sessions can also improve concentration levels as well as increasing input from the 'quieter' members of the group. Those participants not comfortable contributing as part of a large group may be more willing to give their point of view during these smaller sessions. Breakout groups will require more physical space and a possible duplication of resources, but having participants compare and contrast information in smaller groups can lead to all manner of new thinking.

Meeting notes

Not every meeting will have a designated note taker; in fact most meetings rely on individuals to make their own notes, choosing instead to capture only information on resulting actions. This decision is influenced by the meeting type and purpose. However, if a note taker will be in attendance, it is now time to determine the format in which notes should be captured, written up and circulated. Rather than simply relying on handwritten notes, which has the potential to take days to type up and circulate, is there suitable technology you could use to automate this? Chapter 13 gives more information on what technology may be available to assist with the recording of the meeting notes.

reminders

Note taking

If formal meeting notes are to be used it is important that the process is in place and communicated in advance of the meeting.

Individual or group responsibility

Let the participants know in advance of the meeting if a note taker will be in attendance to record the meeting and in what level of detail. This will let the participant prepare any resources for making their individual notes.

Issue a meeting note template

In order to accurately record the meeting information, issue the designated note taker with a meeting note template. This can either be used electronically during the meeting, or recorded by hand, and transcribed immediately afterwards. To download example meeting note templates, visit www.meetingexpert.co.uk.

Brief the note taker

Ensure that the note taker is familiar with any jargon that may be used; this will aid the flow of the meeting. Also ensure that the note taker knows the participants by name – normally contributors will be recorded by their initials for speed of capture.

Collate all relevant material in to a single file

Whether you choose to use technology or not, essentially the notes from your meeting should be circulated in a format that incorporates different types and sources of information as simply as possible. Why? Because too many different files attached to an email are quite daunting to receive and it then becomes increasingly likely that the email will not be opened, let alone read.

Distribute notes promptly after the meeting

The longer the period before participants receive the meeting notes, the higher the probability of forgotten action items and a reduced enthusiasm to advance the project.

Aim to distribute notes within two to three days.

Meeting invitation

Finally, send out the meeting invitation to the participants that you have selected previously. The invitation should include some or all of the following information (dependent on the formality and frequency of the meetings) and should be as appealing as possible.

- Meeting title – to include meeting purpose. If the meeting is of a confidential nature, be careful with the wording of the title as this could be seen in electronic calendars, etc. If sending an electronic invitation use the 'confidential' options available.
- Meeting date.
- Meeting times, start and end.
- Meeting location (include a map/directions if necessary).
- Meeting objectives.
- Meeting agenda.
- Meeting ground rules – circulate the Brilliant Meetings **ACTION PLAN.**
- Details of the other invitees.
- Your contact information, in case of an emergency.
- What facilities are available to those who arrive early, so they can make the best use of their time. Refreshments, Internet, telephone, hot-desk facility, for example.
- What equipment will be available for use during the meeting.
- In what format contributors should bring (or forward in advance) presentations/content etc.
- Dress code (business suit or 'dress down Friday'?).

If this is a new group, or it includes external participants, where possible also include a short biography on individual participants, detailing why their presence is a meaningful addition.

For one-off meetings or the first meeting of a new group, it would also be helpful to incorporate an executive summary – probably written by the meeting leader – as to why this meeting is strategically important to the organisation and to the individuals invited. If participants can identify a clear 'What's in it for me?' they will be more likely to accept the invitation and invest more of their time in preparation.

If there are several weeks between the initial invitation, the draft agenda and the meeting itself, then diarise a reminder to chase up responses and briefing material at least 24 hours before they are due, as well as sending out a reminder about the meeting itself.

Additional preparation

If the meeting is being held off-site or includes people from outside the organisation visiting your site, there may be additional preparation required. Consider the following:

- allocation of car parking spaces;
- visitor passes/name badges;
- security clearance;
- distribute the postcode of the venue for satellite navigation entry;
- send a map and directions, along with public transport information.

brilliant quote

'The best preparation for tomorrow is doing your best today.'

H. Jackson Brown, Junior

CHAPTER 13

In-room meeting resources – if you've got them, use them!

Many organisations have spent millions of pounds on audio visual equipment for meeting rooms, only to watch them gather dust in the corner; the video conference thingy in the cupboard, the interactive whatsit on the wall or the copy board that ran out of paper years ago, etc.

There are many reasons why this equipment is left dormant; the people trained to use it have now left the organisation, the equipment is deemed to be faulty, or maybe you have had a previous bad experience. Whatever the reason, here is the challenge:

> *If you have any of the equipment listed in this chapter available to you, now is the time to find out how to use it (if you do not already know).*

We are not advocating that they should all be used in every meeting – just that by making the best use of the tools already available, you will look good in front of your colleagues, inspire others around you to use the equipment and ultimately save yourself time.

brilliant tip

Make the best use of any in-room equipment and resources already available to you.

In this chapter we offer suggestions for achieving effective use of the most commonly found meeting room equipment/resources available today (listed in order of the most common usage).

Flip chart

Flip chart

A large pad of paper hinged at the top. Normally they are fixed on to a free-standing easel that allows the sheets to be flipped over, presenting the information sequentially.

> The flip chart is the most common in-room meeting resource used today

The flip chart is the most common in-room meeting resource used today, probably because it is simple to use and can be used by everyone. Quite simply a walk-up-and-use resource to note ideas, storyboard thoughts, collate the intellect of the group, or simply present information.

You can prepare flip charts in advance of the meeting, noting down the main points that you plan to cover and asking for feedback which you can then record. If you want to gather information from your audience, use a pencil to add reminder notes at the top of each flip chart page in advance of the meeting. You will be able to see your notes close up, but they will be invisible to your audience. Preparing flip charts in this way can save a considerable amount of time during your presentation.

Dry erase board

brilliant definition

Dry erase board

Also known as a whiteboard, marker board or dry wipe board, a dry erase board is a name for any glossy surface, most commonly coloured white, where markings temporarily adhere to the surface of the board.

Dry erase boards are also a very common sight in meeting rooms. If you have the opportunity, access the meeting room in advance of the meeting and write up bullet points pertinent to your contribution, completing the rest whilst you are delivering your presentation.

Some dry erase boards are magnetic; this presents an opportunity to prepare content in advance and 'post it up' during your presentation.

Data and video projectors

brilliant definition

Projector

A projector displays content from a computer or video source on to a large screen. Projection is by far the most common in-room presentation technology used in business today.

Once the computer is connected to the projector (either by a physical cable or through a wireless interface) images from the computer are then projected on to a large screen, making the information visible to everyone. As well as displaying information from a computer, you also have the ability to show content from a video or DVD player.

When using a projected image to share content, consider the number of participants that will be gathered, together with the room layout, to ensure that everyone has a clear view of the display. Consider also how your information will look when it is presented on a large screen; refer to the presentation guidelines in Chapter 4.

If you are using a portable projector and screen, secure all the trailing cables (for health and safety reasons) and ensure that the projected image is central and square on the screen.

Plasma and LCD displays

Plasma and LCD displays

Plasma and LCD displays are sometimes known as 'thin-panel' displays and can be used to display information from both computer and video sources.

Typically, the size of the image will be smaller than that produced by a projector. However, as technology advances the size of plasma and LCD screens are increasing and they are becoming less and less expensive.

Again, once a plasma or LCD display is connected to a computer or video source, the image is displayed on to the screen for all to see. The same considerations apply to the computer presentation and room set up, as described in the preceding section on 'Data and video projectors'.

Audio conferencing

Audio conferencing

Audio conferencing (also known as teleconferencing or voice conferencing) allows multiple individuals in different locations to connect simultaneously, through conventional telephone services or VoIP (Voice over Internet Protocol).

There are several methods of achieving an audio conference:

- through a dedicated 'conference phone';
- by dialling in to a bridging service that links several phone lines together;
- through a third party organisation, with operator assistance, that links the participants together.

As audio is the only method of communication in this instance, when starting the call remind participants to switch mobile phones off and to minimise ambient noise – such as fingers drumming on tables, etc. There will be a limit on the number of participants that can effectively participate on the call, dependent on the technology being used, and this applies equally to the number of participants that can effectively hear the content of the call.

Video conferencing

Video conferencing

Video conferencing (also known as VC) allows two or more locations to interact via two-way video and audio transmissions simultaneously.

Video conferencing brings people, located at different sites, together for a meeting. In its simplest form VC connects two people in separate locations, known as 'point-to-point'. Larger conferences, where there is a need to connect multiple sites, are typically more complex to set up and are known as 'multi-point'. Besides the audio and visual transmission of meeting activities, video conferencing can be used to share documents, computer-displayed information, and information from an interactive whiteboard.

To ensure that the video conference runs smoothly, a day or two in advance make a test call to the other locations, checking audio levels and camera positioning (saving camera positions if appropriate). Ensure that everyone can see the screen and can be seen by the camera. This might mean a change to the room layout, depending on the size of the group. Remember to organise this for both ends of the meeting (or more if it is a multi-point call).

If possible, try to make sure that the backgrounds behind the participants are not 'busy'.

To ensure that all rooms are ready, arrive 10 minutes before the participants to set up the call and check that the technology is working correctly.

brilliant tip

If you are going to use audio or video conferencing to facilitate a same time, different place meeting, make sure you take into consideration any difference in time zones of remote participants. Your mid-morning meeting might result in the far end participants coming in before their normal start time, or even missing their lunch.

Wireless remote presenter

definition

Wireless remote presenter

A wireless remote presenter allows users to control their computer whilst 'roaming' around the room.

When using a projector to present PowerPoint™ a wireless presenter advances slides remotely, giving the presenter the freedom to engage with their audience and move freely within the room. A very useful, inexpensive tool that really does aid the presenter.

Interactive whiteboard

definition

Interactive whiteboard

An interactive whiteboard is a large interactive surface that is either fixed to a wall, or mounted on a stand. When connected to a computer and projector, an image of the computer's desktop is projected onto the board's surface, which the user can then control using a special pen or their finger.

> the interactive whiteboard is a tool that can add a real 'wow' factor

Although extremely popular and widely used in education, the interactive whiteboard is rarely used in a business context, although it is a tool that can add a real 'wow' factor to training sessions, presentations, collaboration, etc. It gives the user the ability to advance

PowerPoint™ to the next slide simply by touching the board, or you can navigate a Web page using hyperlinks in exactly the same way. With just a little training, presenters can add real-time annotations to any display; perhaps in order to explain a point more fully or in response to questions from the group. These annotations can be captured and either embedded into PowerPoint™, or saved as separate documents for immediate distribution to participants.

If you are going to use an interactive whiteboard in the meeting, ensure that all computers that will be connected have the correct software installed, in advance of the meeting. Again, if you are using a portable projector, ensure that the cables are secured (for health and safety reasons).

Visualiser/document camera

definition

Visualiser/document camera

A visualiser/document camera is an electronic imaging tool that will display a 2D or 3D object (moving or still) on to a large screen.

A visualiser/document camera enables you to place an object under the camera, which is then processed through the computer and displayed via a projector or plasma screen. The smaller units are more commonly known as document cameras. They sit on the desk and will cope with anything up to A4 size. Visualisers can be suspended from the ceiling and may have an inbuilt light source to illuminate the object and can display very small to very large objects.

Response systems

Response system

A response system allows the collation of responses, which are submitted via a 'clicker' or hand-held response pad by the participants in real time, in answer to a set question.

Response systems are not a very common business meeting resource, used mostly in education and training. However, should you have access to a system then they can be used in two ways: to confirm that information has been understood, and to allow for anonymous voting. Such systems will display automatically the percentages of votes cast for each answer option offered. Use the following suggestions to make voting/response sessions most effective:

- Use clear concise wording and phrasing for both the questions and answers.
- Use no more than five options for the answer.
- Ensure that all the voting units have adequate battery life.
- Ensure line of sight between the voting units and the receiver device.

Telepresence

Telepresence

Telepresence can be summed up as high-end video conferencing between two or more small groups, in a specially designed environment that is permanently connected. This is *not* a multi-functional room, instead designed only for telepresence meetings.

Telepresence rooms are designed specifically for conferencing with a small number of participants in each location. If you are lucky enough to have this solution installed in to your environment, then the preparation has already been done. The connection is made, the room configuration is fixed and identical in all locations, and users have been familiarised with the technology.

Summary for preparing a Brilliant Meeting

Define meeting purpose	Identify the basic purpose for holding a meeting. Choose which of the Learn, Share, Create qualities needs to be fulfilled by the meeting.
Physical meeting?	Could a meeting using connecting technologies achieve the outcomes just as well?
Why?	Define the meeting objectives using the SMART acronym and develop the agenda detailing the sequence, responsibility and duration for each item.
Who?	Choose the participants needed to achieve the meeting objectives whilst designating any roles needed in the meeting.
When?	Select the date, time and duration of the meeting, taking into consideration any travel implications for the participants.
Where?	Decide upon the venue most suited to the meeting type and the resources needed to achieve the desired outcomes.
What else?	Reinforce the meeting ground rules, distribute any briefing documentation, schedule catering, breaks and send out the meeting invitation.
Technology	If any technology resources are to be used during the meeting, familiarise yourself with them and check that they are functioning correctly.

Use the checklist to ensure successful preparation for future meetings. Download it from www.meetingexpert.co.uk.

Checklist in Preparation for a Brilliant Meeting

Summary Information
Meeting Title
Name of Group
Date & Time of Meeting
Meeting Venue

space 2 inspire

Meeting Purpose			
Meeting Objectives			
Meeting Title			
Develop Agenda &Timings			
Costs and Benefits			
Choose Participants			
Assign Meeting Roles			
Choose Date, Time & Duration			
Choose Venue			
Book Any Additional Resources			
Distribute Briefing Notes / Previous Meeting Notes			
Collate Presentations			
Meetings Notes			
Schedule Catering and Breaks			
Meeting Invitation			
Send out Agenda & Meeting Ground Rules			
Other			
Last Updated			

PART 3

Next time you lead a Brilliant Meeting

Introduction

You are just about to take to a platform that gives you the ability to demonstrate your preparation skills, facilitation abilities and, most of all, your personal drive for a positive and effective meeting outcome. This will be illustrated through positive engagement from participants, resulting in clear and well-documented actions that ultimately will benefit your organisation. In Part 1, we detailed how participants can use meetings to advance their career; as the leader of a Brilliant Meeting, you now have a more visible platform to raise your own personal profile.

The success of a meeting depends on everyone present working together to their best ability in order to achieve the desired outcomes. Whilst we clearly advocate that ground rules should be in place to ensure that there are no doubts about what is expected and what is unacceptable, it is the leadership qualities that will make or break the meeting.

it is the leadership qualities that will make or break the meeting

A Brilliant Meeting needs a brilliant leader.

There are many different reasons why meetings are held; some will be short and completely focused on a single business outcome, whilst others will have the luxury of time and informality to delve further into the human aspects underpinning the meeting. Some will be a regular gathering of participants well known to each other in familiar surroundings; others will bring together new participants in new surroundings. The former may require virtually nothing in the way of introductions and housekeeping before getting down to the business of the day, the latter will require a very different start to the meeting.

In Part Three many ideas and suggestions have been put forward for use in the meetings that you are leading. But use these with care, as not all of the tools will be appropriate for every type of meeting.

brilliant quote

'The best executive is the one who has sense enough to pick good men to do what he wants done, and self-restraint to keep from meddling with them while they do it.'

Theodore Roosevelt

CHAPTER 14

Setting the tone

First impressions

> Brilliant Meetings need leaders who will quickly assert their influence

Brilliant Meetings need leaders who will quickly assert their influence through professionalism, attention to detail, and strong impartial leadership.

Be early

You need to be in the meeting room before the participants begin arriving to ensure that the room layout is correct, any materials or presentations are ready and, if required, the seating plan is in place. By being early you also have an opportunity – should it be required – to deal with any unexpected occurrence, such as, if the room has been double-booked, incorrect layout, readiness of technology, etc.

brilliant tip

Book yourself a separate meeting to prepare

The time you book the meeting room is usually identical to that advised to participants. To allow for room setup, book a pre-meeting slot for as long as you need before the start of the scheduled meeting, and make the necessary preparations. Now you have access to the room for those preparations and, as far as anyone else is concerned, your calendar shows you as 'busy' and you will not be interrupted.

Start on time

Unless you have advance access to the room, chances are that you still will be preparing whilst some or all of the participants are in the room. Now the meeting will not only start late, but also a vicious circle perpetuates where participants habitually arrive late because experience has taught them to expect a late start.

So make it your business to be in the room well before the meeting start time, have all your preparations made and, at the publicised time, call the meeting to order and begin. Show participants that *your* meetings will start as advised.

Dealing with late arrivals

Most participants try to be in the meeting on time, but you can understand why they might wait until the last minute before leaving their own office, if a culture of late-starting meetings exists. Introducing Brilliant Meetings into an organisation will set the expectation for meetings to now start punctually. Having read Part 1 of this book, participants will recognise that they can use any time in advance of a meeting to their advantage if they arrive early.

If someone is late because of a genuine reason outside of their control, they should at least have the manners to phone and advise you of their lateness. Once the meeting is under way, the 'Apologies' item gives you the opportunity to enquire about unadvised absences. 'Has anyone seen or heard from Duncan?' This sets the expectation that other participants need to contact you should they find themselves running late for a future meeting. Now everyone knows that you expect this same respect from them – Brilliant Ground Rules!

brilliant tip

Encourage on-time arrivals – next time

If a participant arrives late without a genuine reason, give them the task of buying coffee for everyone at the beginning of the next meeting.

If key people are late, you will have to make a decision about whether to start the meeting without them. If you know they are going to be only a few minutes late, a late start will avoid the need to repeat information or change the agenda order. But do you have time to wait and still get through all the business in the allocated time? Either way, inform the participants who are already there what your chosen course of action is.

brilliant tip

Leave a seat closest to the door for late arrivals

If there is no seating plan, ensure that latecomers have seats available closest to the door, which will make their entrance less disruptive to all.

Opening statement

Before you get down to the business of the day and commence with the first agenda item, consider whether this is a meeting that requires you to inform participants of any 'housekeeping' details.

Remember, whatever your first words are, this is your first formal opportunity to address and impress the group as a whole.

- Stand up.
- Thank participants for their attendance.

- Make eye contact with everyone.
- Speak clearly; neither too fast nor too slowly.

Housekeeping issues

Health and safety

If there are any participants for whom the location and group is not familiar, consider what information they need to know. It might include the following:

- fire exits;
- toilets;
- ground rules, especially concerning email, laptops and mobile phones;
- refreshment breaks.

tip

Follow me!

Health and safety is essential for new participants in new surroundings, but typically a little dull and uninspiring.

Make it light-hearted, personal and memorable: 'In the event of the fire alarm, turn left out of the room, and the emergency stairs are 50 yards on the right. If in any doubt, just *follow me* as I won't be hanging around!'

Meeting reminders

Meeting purpose and objectives

Before the personal introductions, which are necessary only if the group members are unfamiliar with each other, reiterate the purpose of the meeting along with any pre-set objectives.

Motivate participants to be engaged in the meeting – give them your vision of what a successful meeting might look like, using emotive, inspiring language.

Confirm that they have the authority to make relevant decisions if this is an anticipated outcome from the meeting.

Agenda timings

If appropriate, reiterate that each agenda item has a time allocated to it, and detail how you will deal with non-completion of any agenda item within that time. A few options are:

- return to it at the end if time permits;
- reschedule other items in this meeting;
- reschedule to a future meeting.

Breaks and refreshments

If you did not mention breaks in your introduction, then restate the scheduled timing of breaks now, and confirm if refreshments are to be provided.

Ground rules

Brilliant Meetings incorporate the **ACTION PLAN** ground rules which clearly state, for all participants, what is expected of them and what is unacceptable. If these ground rules are still new to your organisation it would be worthwhile reminding participants and drawing their attention to the details posted on the meeting room wall.

The Brilliant Meetings **ACTION PLAN** ground rules are detailed in Chapter 2.

Roles and responsibilities

Confirm which participants (if any) have a role during the meeting. For example, note taker, scribe, time keeper, etc.

Working with a new group

If you are working with an established group of participants, then you are ready to start with the first agenda item. If group members are not known to each other, then here are some suggestions for you to consider using so that everyone is clear about how your meeting is being run.

Introductions

Starting with yourself, go around the table asking everyone to introduce themselves. You will have set the expectation with your own introduction, but do not give or expect anything other than a brief introduction unless you have pre-warned participants that something more substantial is required. See Chapter 1, Your personal 'Elevator Pitch' for more guidance.

How discussions are managed

Set out the rules of engagement for discussion contributions made outside of individual presentations. You might insist that anyone wanting to speak must do so by indicating to you, as leader, that they have a contribution to make. This is quite formal, and more likely to be used with larger groups, but it does stop participants from talking over each other, and gives you the ability to spread contributions evenly around the group. You may decide that the discussion is better dealt with by splitting into subgroups which then present their own theories – or alternatively have no formal rules for contributions at all, which works well with smaller groups.

How questions are handled

Quite often questions are raised that would ruin the flow of thoughts and information if they were immediately answered. Inform the group how you intend to deal with any such questions that are deferred, in order to make sure that they are not forgotten. They could be recorded by the note taker (issue a template) or written in full view on a flip chart or whiteboard, so that the questioner knows their question has not been ignored and also, if practical, so that participants can deal with the question in another part of the meeting. If you have external presenters, make these options known to them also. With smaller, more intimate meetings, this approach could be seen as very authoritarian, but for larger departmental or organisational meetings, it almost certainly will be an essential technique for maintaining the flow of the meetings.

Notes

Announce what level of detail will be recorded in the notes, and when and how these will be circulated. This provides participants with more choice about what notes they take for themselves over and above the official version.

Reaching decisions

Not all meetings revolve around making decisions. Meetings are effective ways of exchanging information, announcing future plans, and brainstorming, none of which may require decisions being made.

Meetings are effective ways of exchanging information, announcing future plans, and brainstorming

If your meeting does involve decisions, then is it clear to your participants how those decisions are going to be reached in order to fulfil the meeting objectives and purpose? Basically there are three ways in which a group can be involved in the decision-making process: autocratic, majority rule and consensus.

Autocratic

The meeting will give you the opportunity to take soundings and advice from the participants but, ultimately, you will make the decision. As a business leader it could be part of your management responsibility to reach such decisions. It could also be that you are privy to other strategically confidential or sensitive information not available to your colleagues, in which case only you have the full facts in order to make a fully informed decision.

Majority rule

Once the discussion has finished, the options are stated and votes made in support of each one. The number of votes for, against or undecided are recorded in the notes, and the option with the highest number of votes is selected.

Votes can be cast using a show of hands, secret ballot, or an electronic response system for anonymity.

Consensus

Consensus decision making is a process that seeks a level of support from every participant, and an agreement from those not wholeheartedly in support that they will at least publicly support and implement the decisions reached. Typically, these decisions take much longer to reach as concessions are made to proposals in order to reach the terms that even the least supportive can sign up to. If there is just one participant who will not agree, they effectively have the opportunity to act as a 'blocker' against a consensus being reached.

Whilst consensus is not common in boardroom meetings, it is very relevant for larger groups taking decisions on new working practices, departmental issues, etc.

brilliant tips

When leading a consensus decision-making process

- Review the meaning of consensus and the process of achieving a consensus, then agree on a targeted time period to reach a consensus decision.
- Present the subject and possible outcomes clearly and concisely.
- Encourage all members to present their point of view and allocate enough time for full consideration by the group.
- When discussions reach a stalemate explore the next most acceptable alternative for all parties.
- Distinguish between major objections (which is a fundamental disagreement with the 'core' of the proposal) and possible amendments.
- If agreement happens too quickly and easily, be suspicious, explore the reasons and be sure that everyone accepts the solution for similar reasons.
- Differences of opinion are natural and to be expected, they can actually help the group's decision-making process. The wide range of information and opinions gathered during debates will produce an environment that is more likely to produce acceptable solutions for all.
- Once a decision has been reached, ensure that all the group members feel they have had the opportunity to express fully their individual opinion and that they agree with, and will support, the consensus decision.

CHAPTER 15

During the meeting

Having set the tone, you are now ready to proceed with a Brilliant Meeting in a professional and motivational atmosphere, where participants can give their full attention and contribute honestly and freely in order to achieve the required meeting outcomes. In order to lead a Brilliant Meeting, you may need to take on many varied roles.

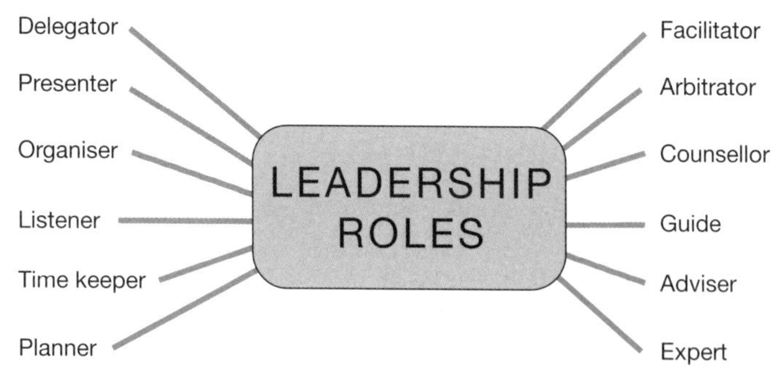

Managing the agenda

tip

Ensure that all participants have a hard copy of the agenda.

Having allocated timings for each agenda item, try to keep to them as closely as possible. The role of time keeper within a meeting can either be incorporated into your role as meeting leader, or it can be delegated to one of the participants.

> The ultimate consequence of poor time management is an incomplete meeting that can damage your credibility

The ultimate consequence of poor time management is an incomplete meeting that can damage your own credibility, and which effectively labels the meeting as 'a waste of time'. You cannot stop a discussion or presentation bluntly mid-flow just because the time allocated has been reached. Agenda items are managed by keeping participants aware of time constraints when necessary, and taking positive actions to keep the meeting on track and to time.

brilliant examples

Keeping the meeting on track and on time

Summarise, conclude and suggest moving on:

'I think we are almost finished on this point, does everyone agree with Duncan's proposal?'

Suggest that this issue is removed from this meeting and 'parked' until the next meeting:

'As this item needs further input and discussion, can I suggest that we defer to a future meeting and allocate an appropriate amount of time to it?'

Suggest that this issue is dealt with by a 'sub group' and presented at a future meeting:

'More detail and discussion is needed than we have time for today. Can I suggest that you three convene a separate meeting and then circulate the briefing notes in order that we can discuss your findings at the next meeting?'

Suggest an extension to the meeting, if the room is available:

'As more detail and discussion is needed than I had originally allocated time for, would everyone be able to stay for a further 45 minutes to complete the meeting?'

tip

Countdown cards

For meetings where formal timed presentations are being given, consider using A4 laminated flashcards with '3 minutes', '2 minutes', and '1 minute' printed on them in large text. Either you or the designated timekeeper can now hold these up to presenters, thereby alerting them to the impending end of their allocated time without directly interrupting their words or the concentration of those listening to them.

Directing the note taker

Chapter 12 provides guidance for the note taker. However, during the course of the meeting it might be necessary for you to specify if a different level of detail should be documented in the meeting notes.

Managing and encouraging contributions

Participants have been selected based on their individual expertise, experience and personality to work together to achieve the meeting objectives. Whether you need to coerce or curtail contributions, getting the best from your individual participants may involve some or all of the following techniques.

Brainstorming

This is a group technique designed to generate a large number of creative ideas for the solution to a problem. At this stage you are collecting ideas to evaluate later and, so long as they are focused on the subject, responses should not be restricted, but instantly recorded without any additional comment. The next stage is to evaluate each response in order to determine which are worthy of further consideration and idea development.

Round robin

To make sure everyone is given the opportunity for individual input, start the discussion by giving everyone, in turn, a precisely defined amount of time to share their initial thoughts with the group. This exercise can be used several times during a meeting so always start off with a different participant, and move around the table clockwise giving everyone their say. You might be surprised to find that there is already a consensus or even unanimity on an issue. More likely you will need to summarise and go back to individual participants to expand on their points.

tip

Congratulate or critique contributions, ignore individuals

The aim here is to disconnect the contribution from the contributor so that, if a suggestion is in need of criticism, the personality can be removed. This is a very difficult skill, bordering on counselling, that would not be appropriate for a high-powered business meeting, but could deliver great value to a meeting with less confident participants.

Identifying non-verbal signs

Whilst the quieter members might not speak up willingly, you should be able to gauge from their body language whether they have something of value to contribute. They might be leaning forward in their seat, or perhaps nodding or shaking their head, in which case you can direct a question to them for their opinion. If they have not spoken out much before, reaffirm their comments in a positive way (remember to congratulate the comment, not the individual).

Assign an action for next time

If you really cannot get a participant to open up, before you consider their value to future meetings, assign them a couple of easy actions that they must report on next time.

'One at a time please'

Whether participants are being argumentative or simply just eager to have their contributions heard, having more that one person speaking at once does not make for a good meeting environment. These simple assertive words will allow you to take control of contributions from a larger group, and suggest some respect is forthcoming from those that are being just plain rude.

Remind participants of the Brilliant Meetings **ACTION PLAN** – 'Imagine the CEO is always present', and ask them to allow speakers to continue without interruption. More tact will be required if the persistent offender is someone senior, as this will be more difficult to request.

brilliant tip

Taking back control from a colleague in a senior position

If a situation arises where you need to regain control, be ready to leap in when a pause occurs, with something along the lines of:

'Thank you, Duncan, that's a great example of one way in which we might proceed in this instance. However, I'd like to get back to the specifics of . . .' or alternatively,

'Thank you, Duncan, that's a great suggestion, but more discussion is needed than we have time for today. Can I suggest that this is added to the agenda for the next meeting?'

This way you are publicly thanking them for their contribution and simultaneously steering the meeting back on track.

If you had been sitting, then stand as you address the situation. If you had been at the side or back of the room, move forward as you speak to claim back your meeting.

If the digressing tendencies of your senior colleague are well known in advance, then you can make arrangements to deal with it; assign the role of time keeper to another participant, and have an arranged sign for requesting them to audibly warn you when the time for the agenda item is almost over. You can then interject, citing this as the reason for wrapping up the disruptive contribution, and move the meeting on.

tip

Questions/comments only via the leader

With smaller, more intimate meetings, this approach could be seen as very authoritarian but, for larger departmental or organisational meetings, it almost certainly will be an essential technique for keeping the meeting flowing.

Questions

Questions are a fabulous method of encouraging information from participants but, for this to be effective, you must practise your questioning techniques and use them where and when appropriate.

There are several great reasons to ask questions of your participants:

- to gain information;

- to check and confirm understanding;
- to arrest current thinking – encourage new thinking;
- to prompt and channel discussions;
- to summarise information.

Individual or group questions

Make it clear if you are posing a question directly to an individual, to the group as a whole, or whether it is open for any member of the group to respond.

- If directed at an individual, start the question with their name:

  'Duncan, how will this new procedure affect your department?'
- If directed at the whole group, maybe you want a show of hands on a proposal:

  'Hands up who thinks we should talk again with the supplier?'
- Or if you simply are looking for a response from the group:

  'Who can tell me what the impact on their department will be?'

Questioning techniques

It is important to understand question types because, when phrased correctly, these will determine the length and detail of the answers given, providing you with the information you were seeking.

The great news is that there are basically only two types of questions to master. An open question generally produces answers with more detail, whilst a closed question generally results in a single word (or at least a shorter) answer.

there are basically only two types of questions to master

Open questions

Ask open questions when you want to 'open up' dialogue and receive longer, more detailed answers. Open questions generally begin with 'what', 'why', 'how', 'who' and 'when'. An open question asks the respondent for his or her knowledge, opinion or feelings. 'Tell me' and 'describe' can also be used in the same way as open questions. Be careful when asking 'why' questions as too many could come across as confrontational.

Here are some examples:

- What happened at the meeting?
- Why did he react that way?
- Tell me what happened next.
- Describe the circumstances in more detail.

Open questions are good for:

- seeking an opinion – ask a subjective question:

  'What do you think about the proposed changes?'
- identifying specific information – ask an objective question:

  'What was the percentage gross margin achieved last year on this product?'
- collating ideas for actions – ask a problem solving question:

'What are your next steps in resolving this recruitment issue?'

Enabling contributions from the quieter members of the group

If you know that a participant has vast experience to contribute, but they are not doing so – ask them a direct question that you know they will be able to answer. It is not the information to the easy question that is of most importance, but getting them conditioned to speaking up now, so that when the difficult questions arise, they are confident about speaking.

Closed questions

A closed question usually begins with: 'are', 'who', 'did', 'will', 'is', 'do', 'can', etc., and you use these to search for short answers, maybe even a single word such as 'yes', 'no', 'don't know' or the factual answer. Of course this will not stop the 'Yes, but you should be . . .' but it will minimise it.

For example:

- 'Are you happy with the meeting agenda?'
- 'Is the room temperature ok for everyone?'
- 'Did you complete the pre-meeting preparation?'

Closed questions are good for:

- testing understanding:
  'So, if I understand this correctly, your group have concluded that the telephone should be answered within three rings, and not four rings as was previously agreed?'
- summarising and concluding a discussion or making a decision:
  'Now we know the facts, are we all agreed this is the right course of action?'
- identification:
  'Who is responsible for reporting back on the actions from the last meeting?'

---

brilliant reminders

- Use the right question – open or closed for short or longer answers.
- Use questions only when appropriate – do not disrupt a full-flowing discussion.
- Keep responses to questions on-track.

- Prepare questions in advance.
- Remain positive when you phrase and ask a question.

brilliant tip

Asking the correct type and style of question makes it easier for the people around you to provide the appropriate answer.

Do not expect to become an instant expert in questioning techniques

Effective questioning is a skill that needs to be nurtured and practised. You can practise these skills in a social environment; try using different types of questions with your family when enquiring about holidays/work/school, etc.

Effective questioning is a skill that needs to be nurtured and practised

brilliant quote

'The uncreative mind can spot wrong answers, but it takes a creative mind to spot wrong questions.'

Anthony Jay

Dealing with challenging behaviour

Most meeting participants focus well on dealing with the topics under discussion. There will, however, be occasions when the behaviour experienced is less than ideal, unhelpful even. Quite often those participants even may be unaware that they are being unhelpful so, in case you come across examples of these behaviours during your meetings, here are some pointers for recognising and dealing with them.

Personalities displayed in meetings

The know-all

The know-all tends to be the most vocal, coercive and predictable and is always full of the 'right' answers. They tend to be a little overpowering and aggressive, quick to offer solutions because of their impatient nature. They will always know best because 'I have done this before'.

Deal with this by thanking them quickly for the benefit of their experience, and 'park' their suggestions, then direct the same question to another participant for their opinion.

The silent agreer

The silent agreer can be identified by almost constant nods of the head, and barely audible 'Mmmm' or 'Yes'. Eye contact and contributions are kept to a minimum, but do not be fooled, the lack of participation could be influenced by a shy disposition, or an unprepared participant.

Deal with this by encouraging contributions from them in a non-confrontational manner. Direct a question to them to which you know they have the answer.

The negative type

The negative type is also loud and opinionated, compelled to tell everyone why these ideas and suggestions are doomed to failure. It is unlikely that they will be open to new ideas, because they are fuelled by a negative attitude.

Deal with this by asking them to propose solutions that help, rather than just stating why 'something will not work'.

The 'off at a tangent' type

The 'off at a tangent type' normally has good intentions, but a meeting can be hijacked with personal anecdotes and by-the-way conversations. They tend to get carried away with PowerPoint™ presentations containing 67 slides, with copious amounts of writing, which they then proceed to read verbatim from the screen.

Deal with this by not letting them deviate from the point, politely but firmly. Try posing a question to the group, asking if they feel these issues are relevant to the present meeting.

The 'whispering' type

The whispering type will frequently hold side conversations during the meeting, which often results in them missing valuable information and distracting other participants. These conversations are often unrelated to the meeting topic, making this behaviour even more disruptive.

Deal with this by aiming non-verbal signals in their direction, such as 'finger over lips'. If this does not work, pause the meeting and reinforce why it is essential that only one person speaks at a time. Another alternative is to ask them to share their comments with the group, but this can result in conflict if they insist on keeping their comments to themselves.

BEHAVE – Use the following guidelines during a meeting if you are faced with challenging behaviour:

Be calm. Do not get involved emotionally yourself.

Engage the individual privately, during a break or generally during the meeting.

Have empathy with the situation.

Ask for a summary from them and listen without interruption.

Verify the root of the problem by asking a question for clarification.

Establish agreement on how to handle the root cause in order to move forward.

Recognising body language

Body language is an important part of communication that can constitute 50 per cent or more of what we communicate. Communication occurs constantly during a meeting. Even if only a small minority of participants are speaking, almost everyone (if not everyone) will be exhibiting body language signals that divulge what they are actually feeling inside.

The more you can recognise and understand how to read the body language signs that your participants are subconsciously communicating to you, the greater the influence you will have over them, to either curtail or bring in contributions. Here are some very basic indicators that will help to further your understanding.

- An erect posture indicates interest and alertness.
- Sitting with arms crossed, across the chest, indicates that the person is putting up a subconscious barrier between themselves and others.
- A 'nodding' participant is in full agreement with you.
- A person wringing their hands is indicating concern.
- A shoulder shrug signals that they do not believe what has been said.
- An excessive amount of leg movement indicates nervousness.

Regular meetings at work can become so routine that we forget the importance of our own body language, and that of the others around us. In addition, if our rapport with colleagues is familiar and friendly, we can very easily lose track of the fact that we are still in a professional setting and should conduct ourselves accordingly.

Dealing with conflict situations

Truly difficult situations rarely arise; however, disagreements and differences in meetings should be viewed as potentially very useful as they can actually drive discussions forward and generate valuable information.

Where differences of opinion are not relevant to the meeting, quick decisive action needs to taken by you to keep the meeting on track. If there are 'hidden-agendas' between participants, then the resolution for these has to be outside of your meeting, and you should inform participants of this directly.

If there is conflict or resistance within the group that is genuinely within the remit of the meeting, then you must now focus on changing negative problematic energy in to positive purposeful energy. The group needs to be engaged and empowered to transform the conflict in to decisive, practical actions. Try these basic guidelines, remembering all the time that you must not get directly or emotionally involved.

- Ask the person with the problem to come forward and write it up. Now with the word(s) clearly displayed for the whole group to see, what suggestions can you draw out to deal with it? Have the person whose perceived problem it was stay and write up the solutions offered. Ask them if there are any suggestions that they now think could work.
- Re-focus the group's thinking on the solutions – not the problems. Continually ask questions such as, 'How can we deal with this?', 'What does a successful solution look like?', 'How will this fulfil our meeting purpose and objectives?'

Interruptions

An interruption can occur for any number of reasons – and with it comes an unwanted break in the meeting which, it has been suggested, can take upwards of 10 minutes to recover fully from. The biggest accidental interruptions can be avoided with planning and management.

- Mobile phones should be turned off.
- If you are not using video or audio conferencing, but have this equipment in the room, have these set to 'Do Not Disturb' modes, and then you cannot be accidentally dialled.
- Use clear signage indicating that a meeting is 'in progress'.

Comfort breaks

This is one of the terms we have adopted – along with 'bio break' – as a euphemism for going to the toilet and, for longer meetings, breaks have already been scheduled into your agenda.

One thing to be aware of is that a participant sitting cross-legged desperate for the toilet will not be a good participant. Knowing they have got at least 20 minutes to wait for a scheduled break can be an anxious time. Some meetings have a 'go-when-you-need-to' policy, for which there are advantages and disadvantages:

- Participants could use it as a cover to leave the room specifically to check email or phone messages.
- A key part of the content could be missed – although you as leader can manage this by asking the contributor to wait until everyone is back in the room or to repeat it.
- Just knowing that they can get up and go to the toilet at any time will put some people at their ease, and thereby increase their effective participation.

- You can announce an unscheduled five-minute 'comfort break' if any of the participants are getting out of hand and you need to diffuse the situation. Maybe the break itself will do the trick, or you can direct them to take five minutes outside of the meeting to sort their problem out – if it is not relevant to your meeting.

You will have to decide how you want to manage this necessary occurrence during prolonged meetings.

Impromptu meetings

Many meetings will occur without advance preparation of an agenda, participant list and presentations, etc., just because of general business events. In these circumstances it is still important for all participants to share and understand the meeting purpose and objectives.

If you do nothing else, take a few minutes at the start of such meetings to consider what you want to achieve, and therefore what needs to be covered. This will give you a meeting structure to follow, and allow you to evaluate the effectiveness of even an impromptu meeting by measuring the outcomes against the initial objectives.

CHAPTER 16

Concluding the meeting

Close on a high!

Just as first impressions are so important, the way you summarise and conclude your Brilliant Meeting will be the lasting impression with which your participants leave. Ensure that you allow sufficient time for a positive, inclusive, 'on-time' close to your meeting. This process alone will result in an effective boost to the group's motivation.

If you have 'parked' questions and issues throughout the meeting it is now essential that these are addressed, or at the very least recorded and added to the agenda for any subsequent meeting. As discussed in Chapter 8, the trap of 'Any Other Business' should be avoided if possible, as this normally creates a meeting within a meeting, usually on items you have less control over. If any such issues are raised whilst you are closing the meeting, suggest that these too are added to the agenda for the next meeting.

An effective, timely closure to the physical gathering of the participants is vital for the good work of the meeting to continue; in many cases the work of the meeting is only just beginning. Participants need to leave the Brilliant Meeting full of enthusiasm and motivation to complete the tasks assigned to them. The following acronym will help you to **CLOSE** on a high:

An effective, timely closure is vital for the good work of the meeting to continue

CLOSE

Confirm and summarise discussions.

List action items and confirm meeting notes.

Outline the meeting evaluation process.

Set date/time/venue for next meeting.

Express your gratitude to everyone for their attendance and participation.

Confirm and summarise discussions

Taking each agenda item in turn, summarise the discussions that took place and where appropriate what decisions were taken. This review process has several benefits as follows:

- ensures that all agenda items have been actioned appropriately;
- confirms understanding of the discussions that took place;
- allows the note taker to check the accuracy of the notes;
- provides the participants with a sense of accomplishment and achievement.

List action items and confirm meeting notes

Following on from the agenda review and, for the same reasons, go over the list of actions that have resulted from the meeting. Taking each action in turn, seek understanding and commitment from the participant involved, taking care to reconfirm and document the timescales necessary to achieve them.

Let your participants know when they can expect to receive the meeting notes and in what format, re-iterating from whom these will originate.

Notes are intended to be an accurate, clear, concise record of the meeting, not a verbatim transcript. If you have delegated production of the notes to someone else, then these should be signed off by you as an accurate record of the meeting. Ideally, notes should be circulated within two to three days in order to maintain momentum, and in a file format that cannot be edited by recipients. Chapter 12 gives more detail on meeting notes.

Notes are intended to be an accurate, clear, concise record of the meeting

At the very least, notes should include the following information:

- name of group/name of meeting;
- date and venue of meeting;
- meeting purpose;
- participants;
- apologies;
- agenda items with resulting actions;
- date of next meeting.

We receive masses of information clamouring for our attention on a daily basis. However, it is essential that meeting notes are read, giving participants a timely written reminder of their agreed actions, rather than a last-minute reminder prior to the next meeting. Increase your chances of success by using a format that is clear to read, setting out the information in an accessible way.

Brilliant example of meeting notes

Space 2 Inspire Limited

Strategy Meeting at Central Boulevard Oxford

13 May 2009

FY08 Overview and 3 Year Plan Development

Participants	Apologies
Duncan Peberdy	None (who would miss a Brilliant Meeting!)
Jane Hammersley	
Debbie Maitland	
Hugh Davies	
Rob Boogaard	
Vicki Hing	
Sophie Peberdy	

Agenda item	Summary of discussion and action	Action by	Completed by
1	**Notes circulated after last meeting** Record here when the notes were circulated and that, by a show of hands, they were accepted at this meeting as being a true account of the last meeting.	DLP	13/05/09
2	**Accounts, discussion and review by geographical location**		
2.1	North and Midlands Appropriate details and actions, recorded here.	JLH	17/07/09

2.2	London		
	Appropriate details and actions, recorded here.	HKD	17/07/09
2.3	South		
	Appropriate details and actions, recorded here.	DLP	17/07/09
2.4	Mainland Europe		
	Appropriate details and actions, recorded here.	RBB	17/07/09
2.5	Scotland, Ireland, Isle of Man		
	Appropriate details and actions, recorded here.	DEM	17/07/09
2.6	United Arab Emirates, Doha		
	Appropriate details and actions, recorded here.	TVH	17/07/09
3	**Strategy for new business**		
3.1	Appropriate details and comments, conclusions, etc., recorded here.	JLH	08/06/09
3.2	Appropriate details and comments, conclusions, etc., recorded here.	DLP	08/06/09
4	**New business pipeline**		
4.1	Details of all upcoming projects and the percentage likelihood of winning the business.	ALL	17/07/09
5	**Marketing**		
5.1	Information here on the presentation of a proposed new marketing campaign with costs, timescales, target audience, etc.	VH	29/05/09
6	**Items carried over or added to next meeting agenda**		
6.1	Matters arising or those items that need further research and discussion are noted here.		

Date of next meeting

11 November 2009

Signed by D L Peberdy

Date 14 May 2009

Sample Notes Layout © Jon Moon, *How To Make An Impact* – FT Prentice Hall

brilliant impact

Personalise meeting notes

When we scan a set of meeting notes, we instinctively seek out from the entire document only the agenda items and actions that have our initials assigned to them. If possible, add a section towards the top of the notes that pulls together all the actions for individual participants in one place.

It does require additional work as you have to produce a set of notes tailored for each participant – but they will be the best Brilliant Meeting notes participants have ever received.

Outline the meeting evaluation process

Although this may not be a necessary action after every meeting, it is important, periodically, to collate any suggestions that your participants may have that will improve their experience in forthcoming meetings.

There are many evaluation techniques available depending on the depth of responses you are seeking. However, the golden rule is not to ask *closed* questions – those questions that require only a *yes/no* response, or a show of hands. The feedback from this would frankly be a waste of time.

brilliant example

When dining in a restaurant, how many times has the manager come to your table and asked, 'How was your meal?' Most people instinctively respond with 'fine' or 'great' or maybe even 'fantastic', but does this superficial feedback really help the restaurant to make improvements?

What if his question was phrased differently – 'Excuse me, I am the restaurant manager and I was just wondering what we could have done differently to make your experience with us better?' This question will provoke a considered response, with valuable feedback, not exclusively concerned with the quality of the food alone.

Instead, ask for responses through the use of direct, open questions designed to measure specific aspects of the meeting: venue, administration, inputs, time, etc. These questions can either be issued in paper form, or simply be put to *all* the participants during the meeting and receive verbal feedback that is instantly captured on a flip chart or whiteboard. This does not have to take a long time, but it works well at the end of the meeting to include feedback from everyone.

brilliant examples

Ask participants to start their *positive* feedback with the following phrases:

- 'Something that worked well was . . .'
- 'I really liked the way . . .'
- 'It was a great idea to . . .'
- 'What I liked most about the meeting was . . .'
- 'I think the major reasons for success were . . .'

Ask participants to start their *formative* feedback with the following phrases:

- 'Perhaps during the next meeting we should try . . .'
- 'I would like to cover . . . aspects in more/less detail.'
- 'One thing I was not too comfortable with, was . . . because . . .'
- 'I noticed that the meeting went 'off course' when . . .'

Another way to collate feedback is to use an electronic response system. These systems consist of individual voting handsets that are issued to each participant, with which they respond to questions posed by you. These responses are then collated and can be presented instantly in a graphical format that is visible to everyone.

As an alternative, consider using the services of a Web-based questionnaire for feedback. There are numerous companies that offer this service for a modest sum (some are even free of charge) and participants particularly like them because of their anonymity. The downside is that it takes a little time to set up, but typically anonymous responses are more measured and are more considered.

Set next meeting date/time/venue

If the meeting is one in a series, confirm when the next meeting will take place and establish if any of the participants have conflicting commitments that will prevent them from attending. If, for example, a couple of participants are on annual leave, you have the opportunity for a quick 'round table discussion' on alternative dates and times whilst everyone is present.

Express your gratitude to everyone for their attendance and participation

Finally, conclude the meeting formally by thanking the participants for their attendance and input. You can also take the opportunity to remind them of the next steps: their individual commitment to any actions and confirm the date of the next meeting. It is worth giving a special mention to those who have travelled extensively to be there, those who have undertaken research for a project, and to those who have delivered presentations or authored reports.

If the completed meeting ends before the scheduled time, an early close is more effective than trying to pad the meeting out.

CHAPTER 17

Same time – different place

The frequency of real-time virtual meetings will increase, as instant and effective communication within and between companies drives the use of viable alternatives to face-to-face meetings. Escalating fuel costs, a growing emphasis on corporate social responsibility towards carbon footprint reduction, and the opportunity to save on infrastructure costs, have increased worldwide sales of virtual meeting solutions significantly; this will lead increasingly to virtual meetings encountering issues associated with culture, language and time.

Irrespective of reason, a meeting has been scheduled that involves participants in different locations meeting at the same time. Technology is available to enable this to happen effectively and your job as meeting leader is to facilitate this meeting as you would any other. However, for a multi-location meeting, there are a few aspects that you will need to manage differently to ensure a Brilliant Meeting outcome.

In Part Four we consider the measures organisations should take to encourage use of connecting technologies; issues around the best room environments and the culture within the organisation. But for now, if you are leading a virtual meeting, there are only a small number of variables in your control that can have a positive impact on these meetings.

there are only a small number of variables in your control that can have a positive impact on these meetings

Essentially there are four types of virtual meetings.

Video conferencing

Video conferencing (also known as VC) allows two or more locations to interact via two-way video and audio transmissions simultaneously.

Audio conferencing

Audio conferencing (also known as teleconferencing or voice conferencing) allows multiple individuals in different locations to connect simultaneously, through conventional telephone services or Voice over Internet Protocol (VoIP).

Data conferencing

Data conferencing is a communication session between two or more participants sharing computer data in real time with the ability to see and control each other's computer.

Web conference

A Web conference is a method of conducting live meetings or presentations over the Internet.

Multi-location, but one leader

Regardless of how many locations are connected, there should still be only one meeting leader

Regardless of how many locations are connected, there should still be only one meeting leader. One person, irrespective of where they are located, who controls the meeting, manages participation, etc.

Additional set-up time to include the testing of equipment should be scheduled into your Brilliant Virtual Meeting, and be prepared for the meeting to take longer due to necessary interactions associated with the use of the technology.

Just because the meeting is being held in two or more physical locations, does not mean that any of the ground rules or common meeting etiquettes should be suspended.

Any side conversations will be picked up by the microphones, which will be heard just as well as the main speaker's voice in the connected locations. Likewise, any nervous or hyperactive finger-drumming on the table will come through loud and clear. One way to minimise this is to use the remote control to mute the microphones when the other location is speaking but, better still, to discourage such behaviour.

What time is it?

So you've encouraged the participants, by providing bagels and coffee, to arrive early for a breakfast meeting, in order to connect to your factory outside Mumbai. 7.30 a.m. your time is 1.00 p.m. in Mumbai – you are all sitting there with breakfast, whilst in Mumbai it is almost lunch time and nobody thought to provide colleagues in Mumbai with refreshments. Whilst this should have been considered as part of the planning, if there is time available, suggest they call and have someone bring refreshments into their room.

Introductions

Virtual meetings lose the benefit of pre-meeting social interaction, from exchanging greetings and pleasantries over coffee before commencing. Once everyone is present in both locations, direct participants to introduce themselves, thereby building up cohesion through social networking whilst simultaneously getting accustomed to using the technology, for speaking and listening.

Identify the speaker

When audio is the only medium being used, or when there are more than just a few people for video conferencing, the person talking should always be introduced or state their own name every time they make a fresh contribution. This way it is very clear who is talking, enabling the connected locations to address any questions correctly and also the person taking the meeting notes knows exactly who made the comment.

Body language

In a telephone conference body language is lost altogether; no visible raising of eyebrows or shrugging of shoulders. But, in dedicated telepresence rooms – top-of-the-range video conferencing installations where all the participants are clearly visible simultaneously to each other – all facial expressions and body language are seen as if in the same room. Where many participants are gathered in a room sharing one video camera and screen, it can be more difficult to see the body language of everyone. Some systems are programmed so that the camera automatically zooms in on the speaker, leaving the other participants invisible to their distant colleagues.

Where body movements would normally be used in a meeting as a silent contribution, a nod or shake of the head, thumbs up, etc., remind participants that such signals must be replaced with words.

Mother tongue

If your virtual meeting includes participants that have different first languages to each other, then encourage the use of speech that is clear, concise and spoken at a moderate pace, and void of slang or references to local issues.

Cultural differences

brilliant example

In a formal business setting, an Afghan man would consider it most impolite to make eye contact with a female. If the female was unaware of this, she might interpret this body language to be a result of shyness at best, or discrimination at worst, when neither is actually the case.

Whilst this is an extreme example, be aware of other cultural and religious differences that could make a difference to the way that participants interact, and ultimately endanger achievement of the desired outcomes.

Managing participation

Use round-robin questioning to get input from everyone; rather than have input from participants in one location before moving to the next, alternate between locations, and call participants by name for absolute clarity.

If there are any silences during a telephone conference, either explain to the connected locations why this is (because participants in your location are consulting technical manuals, for example), or ask this question of the other locations. Silences can be interpreted in many positive or negative ways, so clarification is always best.

Sharing content and presentations

If you want to share computer information during a video meeting it is essential to determine what data formats are supported by all the systems being connected, and prepare accordingly.

Meeting evaluation

Telephone and video meetings can be prone to misunderstandings and, given that very little training, if any, takes place in using these technologies, many participants do not feel comfortable or adequately prepared. As part of your summing-up of the meeting, use a quick round-robin to ask everyone how the experience was for them. A more thorough evaluation can be carried out later, but this will give you a good feel for the suitability of a 'same time, different place' meeting.

CHAPTER 18

After the meeting

Discussions were productive and good decisions have resulted. Now to maintain the credibility of the meeting, all the agreed actions must take place and the decisions implemented.

to maintain the credibility of the meeting, all the agreed actions must take place and the decisions implemented

Your influence on these results is now focused on producing a clear review of the meeting in the form of meeting notes, circulated to participants within 72 hours of the meeting concluding, in order to maintain the group's enthusiasm and commitments to their individual actions. If the role of note taker was delegated to another participant, ask them to write up the notes and send them to you in electronic format. Collate together other relevant material used in the meeting, presentations, spreadsheets, etc., and amalgamate them into one easy-to-view file. It is quite normal now for such documents to be sent electronically.

brilliant tip

Print out your Brilliant Meeting notes and distribute them to senior managers with a stakeholding in the decision, along with your direct line manager.

Your direct role in the 'meeting cycle' is almost over for now, as the responsibility shifts from you to the participants who have committed to completing actions, implementing a decision, or signed up to a new way of working.

Before you start working on your own actions from the meeting and developing the agenda for the next meeting, there are still a couple of very important post-meeting issues to complete.

Evaluate meeting feedback

If you took the opportunity to gain feedback from your participants, evaluate this as soon after the meeting as possible. Decide on whether this feedback is to be shared with the group at the start of the next meeting but, most importantly, implement the good suggestions and change the format or style of your meeting, as appropriate for the group.

tip

Remember to complete the meeting evaluation yourself, and be honest!

Managing follow-up actions

There is a very thin line between interfering and supporting participants in the completion of actions that they have either undertaken or been assigned. So long as they have the authority and necessary resources to complete their actions, they should now be left to get on with it.

However, there are strategies that we can use to encourage and assist in the completion of those tasks.

- In the body of the email with the meeting notes, remind participants that actions are listed and need to be completed to the timescales detailed – then re-affirm the date of the next meeting to really focus the mind.

- A week before the next meeting, re-issue the meeting notes again, together with the agenda for the next meeting.
- When issuing the agenda for the next meeting, line list a review of the last meeting's actions, inserting the name of the individual responsible.
- If participants are in the same office as you, suggest a quick catch up over coffee.

Wider communication

Outside of those directly involved in the meeting itself, who else needs to know what decisions have been reached, what actions are being taken, or where the project now stands in the overall timeframe?

In addition to senior management and executive sponsors, findings and progress updates may also need to be communicated to a regulatory body or a government agency. This communication outside of the meeting is a key element in the overall success of the group. It is a great feeling for participants when other people know that they are doing a good job and making real progress.

It is a great feeling for participants when other people know that they are doing a good job and making real progress

Either as part of your own remit or working with the participants, you need to determine what exactly is going to be communicated, to whom it is going to be communicated, and when and how to send out the communication.

The 'how' is perhaps the most important, as it can have a major impact on how the news is interpreted. A simple announcement delivered by email to all employees will be void of any emotion, instead, effectively delivering only cold hard facts and figures.

CHAPTER 19

Do something different

Given that so much of the working week is spent in meetings, it is all too easy for them to become habitual. In a fast-moving business environment with deadlines to meet and clients to satisfy, there is rarely the luxury of time to permit anything other than the necessary business of the meeting itself.

But not all meetings are like this and, occasionally, longer meetings, such as staff training weekends and annual sales kick-off events, allow more time for activities not directly associated with business objectives.

So, given more time and less need to absolutely focus on the business specifics, think about introducing some new characteristics into these meetings that will stimulate, energise and motivate participants in to producing better outcomes.

think about introducing some new characteristics into these meetings that will stimulate, energise and motivate participants

In this chapter there are some simple, effective, yet inexpensive, ways in which you can introduce an element of uniqueness into the meetings that you lead. Whilst they are not suitable for every environment, if just one of them is suitable for inclusion in your meeting, it will remain in the participants' memories.

Just be aware that in any situation when the unexpected happens, people can find themselves outside of their comfort zones. Whilst none of our suggestions is designed to have this effect, you need to think carefully about introducing such innovation into your meetings and, if appropriate, give participants advance notification.

Aroma

People recall smells with a 65 per cent accuracy after a year, while the visual recall of photos sinks to about 50 per cent after only three months.

www.senseofsmell.org

Our odour memories frequently have strong emotional qualities and are associated with the good or bad experiences in which they occurred. Olfaction is handled by the same part of the brain (the limbic system) that handles memories and emotions. Therefore, we often find that we can immediately recognise and respond to smells from childhood such as the smell of clean sheets, cookies baking in the oven, the smell of new books or a musty room in Grandma's house. Very often we cannot put a name to these odours yet they have a strong emotive association even if they cannot be specifically identified.

This material was reproduced from senseofsmell.org with permission of SOSI.

Imagine being able to boost the productivity of your meeting through the introduction of a subtle aroma into the meeting environment. Research has shown that the smell of jasmine or eucalyptus boosts productivity and helps to prevent drowsiness. More specifically, when the scent of jasmine was introduced into a work environment, keyboard errors were reduced by almost 30 per cent and this reduction in mistakes was increased to 50 per cent when the smell of lemon was introduced.

Music

Most of us have a musical soundtrack to our lives; specific songs and tunes that evoke strong memories of people, places and times in our lives. Music can make us happy, thoughtful or sad. How would your participants feel if they entered the meeting room to a piece of music? Instead of fumbling around for small talk as everyone arrives, what would be the effect of a timeless pop song or rousing classical piece? Would such music help to lift moods and attitudes?

Music has the following benefits:

- It helps us to retain messages – researchers have found that it stimulates the brain responsible for language and memory. Think about how you learnt your ABCs as a small child – can you still remember the tune and, more importantly, the alphabet?
- It helps us to make connections – because music utilises both sides of our brains it allows us to connect multiple neural pathways between the left and right sides of the brain, so making connections to memories.
- It can affect your mood by triggering the neurotransmitter 'serotonin' which influences attention, learning and mood. It can also affect the hormone 'epinephrine' which is part of our 'fight or flight' response.
- It can motivate – music can be inspiring and uplifting, for example national anthems should be inspirational and create a sense of belonging for their citizens. Military forces also harness the power of music, setting marching drills to songs, which also builds camaraderie.

brilliant tip

For a group that meets regularly together, you could ask each participant in turn to choose a track that will be played at the beginning of each meeting. Maybe that person could take a couple of minutes to explain why they chose it? It may seem frivolous to use up five minutes of valuable time with a piece of music, but it allows participants to settle quickly in to listening mode. Taking this one step further, you might consider asking everyone to 'check in to the meeting' for the same reason.

Personal check-ins

Before getting down to the business of the meeting itself, give each participant a couple of minutes to talk about where they are at personally. Did they have any stress getting to the meeting; last-minute changes to childcare arrangements, transport that was late, traffic problems, etc. Do they want to share anything about their personal life with the group, how their children are doing at school or sport, what holiday they have booked for this year, etc.

Is there a business advantage to be had from such a seemingly personal invasion? When this process is used for the first couple of times, there may be those who feel uncomfortable with it but, with careful management and objectives, it can produce tremendous value.

- Each person has the opportunity to speak about something completely within their control. Having already spoken at the meeting, when it comes to the business input, they are more likely to contribute further.
- Everyone is listening. If we are dealing with everyday business processes then it is all too easy to switch off from intense listening. But we will listen intently to the personal

contributions, and this puts participants into listening mode too.

- Participants can identify any common ground they share outside of the business environment. These similar personal experiences can draw them together, thereby strengthening the group through these understandings and consequently producing better group outcomes.

If your organisation currently does not start meetings with a 'check-in', then it will need careful introduction in order to become a valuable business ethos. It must be seen to be supported by senior management, and participants should be alerted prior to its introduction; what the expectations are, how it will be managed, and the personal and business benefits for it.

Ice breakers

Ice breakers can be an incredibly powerful tool to create environments conducive to group working by breaking down any barriers that inhibit collaboration. As such, these exercises should be properly scheduled and managed to ensure they do not eat in to the precious time allocated to other agenda items. Establishing some commonalities with fellow participants will help to create a warm, friendly, personal environment, where the participants will feel able to *learn*, *share* and *create*.

Traditionally we expect 'ice breakers' to be used at the outset of a meeting. However, consider using them as 'thinking interrupters' to inspire a group or to get people moving around the room, and re-energised. They can be used as 'lead-ins' for particular topics, or inspiring group sessions used to re-align the group after a break. They can be fun, amusing, humorous, thoughtful, surprising or just plain silly. Here are 10 examples to get you started, and you could always create your own.

- **Inspirational quotes**
  Encourage every participant to bring along a quote that is personally inspiring for them, which they can share at the outset of the meeting and explain its significance. This will also provide a personal insight in to the motivational factors of each participant.

quote

'Satisfaction does not come with achievement, but with effort. Full effort is full victory.'

Mahatma Gandhi

- **Digressing Duncan and Jovial Jane**
  Ask each participant to choose an adjective that begins with the first letter of their first name and one that really matches their personality. Have them introduce themselves just as they wrote it on the card and allow time for others to ask questions. Allocate just 30 seconds per participant.
- **Birthday partner**
  Have participants mingle in the group and identify the person whose birthday (not year – just month and date) is closest to their own. Find out two things they have in common, then share this with the group.
- **Roam and stick**
  Line the walls of the room with different problems for group members to solve, beginning with phrases such as, 'How can we . . .' 'What would the . . .' (e.g., 'How can we reduce the response time for call outs?') Give each group member a pack of Post-it™ notes and invite them to walk around the room, write ideas on their notes, and stick them on the problem to which they apply. They should also be encouraged to discuss with other participants whilst walking around the room.

- **An opportunity to learn**
  Designate a portion of each meeting for 'meeting skills training'. Introduce a new skill at each meeting, such as note taking, presentations, communication, etc.
- **Here we go again!**
  Ask the group to identify the most common types of disruptive behaviours in meetings (interrupters, off on a tangent, manipulators, side conversations, 'that will never work', etc.) Anytime someone exhibits one of the disruptive behaviours, any group member (or the whole group) can shout out 'Here We Go Again!' to gently remind the 'offender'.
- **Fill in the blanks**
  Ask participants to share one or two 'burning questions' they hope will be answered in this session. Write these on to a flip chart or whiteboard, making them visible throughout the meeting as a constant reminder of individual objectives.
- **Sunshine**
  Everyone writes their name in the centre of a piece of paper and draws a sun around it. The paper is then passed to the right, where the next person will write something positive on it. This continues until the paper is back to the person whose name appears in the sun with lots of positive messages all around.
- **Line up**
  As people enter the meeting, hand each one a piece of paper with a different number written on it. Ask the group to arrange themselves in numeric order without using their voices, hands or showing their number. This helps the group to think of other ways to communicate with each other and to work together to achieve a common goal.
- **Name badges**
  Prepare name badges for each person and put them in a box. As people walk into the room, each person picks a

name badge – not their own. When everyone is present, participants are told to find the person whose name badge they drew, introduce themselves and say a few interesting things about themselves. When everyone has their own name badge, each person in the group will introduce the person whose name badge they were initially given, and mention something of interest about that person. This helps participants get to know and remember each other.

Exercise and nutrition

Stand-up meetings

For meetings scheduled to last less than 30 minutes, consider the value of hosting a stand-up meeting. When we sit we have the opportunity to lean back in our chairs, fold our arms, and switch off; in fact how many times have you seen colleagues momentarily doze off during meetings? These situations could not have occurred in a stand-up meeting. Naturally the furniture to support such meetings must be correct, and the group must not be too large; we would recommend no more than the leader plus six others.

Many office environments have areas furnished with high tables and stools – just remove the stools for however long you need. Typically you would not use such an environment for a long meeting, but you could start the meeting without the stools and re-introduce them after half an hour if necessary.

By standing up, everybody is more energised and will concentrate for longer

By standing up, everybody is more energised and will concentrate for longer. Some people have electric height-adjustable desks. These are used to break up the day allowing change from standing to sitting, and for those with back problems, standing for some of the time is a real bonus. For very small meetings of two or three people, you can use such desks to host them as stand-ups.

As a footnote, you need to be conscious of any participants who may not be able to stand for such a length of time. For shorter meetings of less than ten minutes, stand-up meetings can be used without any furniture at all. In fact, given a nice sunny day, they can be held outside to great effect.

Anyone for a massage?

Hire a professional masseuse to give a five-minute neck massage or foot rub to participants during the meeting. This could improve concentration, motivation and even reduce stress levels during a meeting. In addition to virtually guaranteeing attendance, this can be used to start and end the meeting on time. Anyone arriving late forfeits their massage.

---

brilliant example

In 2007 an article in *World Business* reported the following:

'Foot massages are a part of office protocol in Procter & Gamble's Manila office. The new P & G Philippines President, Mr. James "Jim" M. Lafferty, has implemented a new policy which might just revolutionize the concept of worker benefits. Every Tuesday and Thursday, employees and guests who care to, may now have free foot massages during meetings at the P & G offices in Ayala Avenue's 6750 Building.

'Although some (people) were initially reluctant to remove their shoes in front of their colleagues in the middle of a presentation, nobody misses meetings now, according to Mr. Rafferty.'

Nutrition

Biscuits and cakes are great for meetings as they tend to appear magically at tea and coffee breaks. A chocolate-covered sugary biscuit will result in a short sharp rise in blood sugar level, but the effect will not be prolonged. The reality is that we have become programmed to having biscuits and cakes, and the prospect of fresh fruit and cereal bars can seem like a punishment. But these healthier snacks will produce a longer injection of energy, and most of us know that plain water is better for the body and the brain than caffeine-based drinks.

Flip chart fun!

- Prepare a flip chart with a list of your presentation headings. Locate this at the back of the room, as a prompt for delivering your content without the need for hand-held notes.
- If you are looking for 'compare and contrast' information from your audience, use two flip charts next to each other to collate the pros and cons.
- *Touch*, *turn* and *talk*. When using your flip charts, write down the information, turn to your audience, and only then talk about the point.

The attention grabber

example

Display complete faith in your product or service

Many years ago, whilst sitting (and trying to stay awake) through yet another product demonstration, the manufacturer's representative suddenly climbed on to a table and dropped his new (and very expensive) product – a Canon Ixus Digital Camera – from a great height. I suspect he noticed the waning interest of his audience, and immediately took this extreme action to win back our interest and confidence in the product. Well guess what – it worked. More than seven years later, I remember the product, the manufacturer and the name of the representative, and have recently purchased my second Ixus camera!

Breakout groups

A breakout group is a term used to describe the division of the main gathering of people into smaller sub-groups. Each sub-group is given a specific task, with a set of objectives associated, that needs to be considered, within a given time period. This task can be the same for all breakout groups, or they can all be assigned different tasks and objectives. After the time period has elapsed each sub-group will be given the opportunity to present their findings to the main group.

A breakout group may range in size from 2 to no more than 10 people. The only real constraints associated with the number of breakout groups are the availability of the correct facilities, and the ability to facilitate the discussions once all the groups return.

Breakout groups have several advantages:

- They allow participants to actually reach their own conclusions, in an interactive and inclusive environment where everyone has the opportunity to contribute to the discussion. This is especially valuable with some participants who may be reluctant to express themselves in a large group.
- They get people moving around the room and talking to different participants; maybe even sharing new opinions and ideas.
- Work can be accomplished much more quickly with a small group than trying to do it with a large group, especially if

there is a lot of content to be covered. Each breakout group can cover a separate question or issue; once findings are shared, the collective group can agree actions and timescales for implementation.

- Breakout groups can be used to change the group dynamics during the meeting, effectively mixing things up to keep everyone fresh and boost concentration levels.

Take away

a positive way to close the meeting is to ask each participant to verbalise their most important take away

Whilst time is generally limited in meetings, a positive and uplifting way to close the meeting is to ask each participant in turn to verbalise their most important take away from the meeting. What is the one thing that struck them most about what they *learnt*, what was *shared* or what was *created* from the meeting?

Summary for leading a Brilliant Meeting

First impressions	Arrive early and prepare presentations/technology. Start the meeting on time. Deal with late-comers and unexpected events.
Working with a new group	Outline housekeeping issues and explain ground rules. Are decisions being made? If so, how will they be reached? Have participants been assigned any meeting functions, i.e. timekeeper?
Managing the agenda	Keep the meeting on track and on time. Manage and encourage contributions. Ensure effective use of questions. Deal with unhelpful individuals, conflict and interruptions.
Concluding the meeting	Summarise discussions and actions assigned. Confirm when notes will be circulated and if the meeting is to be evaluated. Schedule next meeting. Thank everyone.
Leading a virtual meeting	Test technology in advance. What is the time of day in the other locations? Manage participation between locations. Be aware of any cultural or language differences.
After the meeting	Evaluate meeting feedback. Ensure that participants complete their actions. Ensure wider communication of meeting outcomes.
Do something different	Use aroma and music in meetings. Discover personalities through 'personal check-ins', 'ice breakers' and 'break-out' groups. Take advantage of introducing exercise and nutrition. Grab their attention! Ask for people's take aways.

Checklist in Preparation for a Brilliant Meeting

Summary Information
Meeting Title
Name of Group
Date & Time of Meeting
Meeting Venue

space 2 inspire

Meeting Purpose			
Meeting Objectives			
Meeting Title			
Develop Agenda &Timings			
Costs and Benefits			
Choose Participants			
Assign Meeting Roles			
Choose Date, Time & Duration			
Choose Venue			
Book Any Additional Resources			
Distribute Briefing Notes / Previous Meeting Notes			
Collate Presentations			
Meetings Notes			
Schedule Catering and Breaks			
Meeting Invitation			
Send out Agenda & Meeting Ground Rules			
Other			
Last Updated			

PART 4

A Brilliant Meeting every time

Introduction

The majority of this book is aimed solely at providing practical advice and ideas for the people who organise, attend and lead meetings. Like a road map, following the Brilliant Meetings guidelines will show the route, with possible alternatives, to reach your destination and, like a road map, it will always be there for reference and confirmation. We do not want to add to the many theories and ways of working already promoted by so many books and 'authorities', but fundamentally want to demonstrate to everyone that productive, effective meetings can be achieved realistically as the norm, not the exception.

Having experienced poorly organised and badly run meetings over many years, we realised that the first measures needed to start transforming meetings into positive experiences are very simple and straightforward to introduce. In fact it is the people attending meetings who will make those first small differences that ultimately will result in much improved experiences for all.

CHAPTER 20

The organisational benefits

Do bad meetings really exist?

The senior executives who have the responsibility of introducing improvements into organisations – the ones who determine whether investment is needed in technology, facilities and cultural change – are themselves unlikely to experience bad meetings due to their very position within the organisation. They have the respect of people around them; they have the support of staff to organise, prepare and record every meeting, and they have the meeting room facilities on the executive floor which are not double-booked, are not interrupted, and which are correctly prepared and always ready on time.

So for those of you in senior positions reading this book, do you know:

- how effective meetings are elsewhere in your organisation?
- how your employees, suppliers and customers rate the productivity of meetings held within your organisation?
- what impression your typical meeting environments convey?
- the true utilisation of your meeting spaces?

If you are unsure what the answers would be in your organisation, then perhaps your first call to action is to undertake a 'Meeting Effectiveness' audit. Just be sure that you conduct this survey in such a way that employees tell you what you need to know, rather than what they think you want to hear. Perhaps consider enlisting help from an independent organisation to undertake such a survey on your behalf.

How to effect a change to a positive meeting experience

In today's business climate an increasing amount of work is produced through collaboration, which means that meetings and meeting spaces are actually on the increase. So now is the time to ensure that meetings are conducted for maximum benefit.

Whilst it is impossible to cover every meeting eventuality, Brilliant Meetings is designed to be a commonsense, practical and workable approach to improving meeting effectiveness. Individually, meeting participants, organisers and leaders are not capable of bringing the benefits of improved meetings to the whole organisation. The tipping point of additional benefits such as competitive advantage, increased productivity, reducing time to market, etc. will be truly realised once the organisation adopts the guidelines outlined in this book as an intrinsic part of their culture.

Begin improving meeting effectiveness by asking the following four questions of each and *every* meeting:

- Is the meeting purpose clear?
- Is a physical meeting necessary?
- Are all appropriate preparations being made?
- Will it be effective in delivering the pre-determined objectives?

Understanding why people meet

Essentially there are three categories of meetings, broken down by the basic meeting purpose – learn, share, create.

The *Learn Meetings* are generally when a group gets together to be trained or hear information that is being disseminated to them. This is a less interactive type of meeting: an organisational meeting, a training course, a budget meeting, etc., but nevertheless the best and most productive use of each participant's time is the ultimate business goal.

The *Share Meetings* typically are scheduled on a regular basis in order to facilitate knowledge exchange: sales meetings, meetings of departmental heads, the one-on-one meetings, staff meetings, board meetings, process reviews, etc. Generally we know what to expect from these meetings as they become part of our everyday routine and are the life blood of organisations.

Create Meetings can still take place regularly but the format and participants are subject to change as the process or challenge that the group is tackling evolves. These meetings are more prone to failure because of their ad hoc and less formal nature, although the desired outcome is likely to be a decision, process or product.

Do effective meetings matter?

Effective meetings:

- achieve the same outcomes from shorter meetings, or improved outcomes from the same length of meeting;
- improve employee, customer and client experiences;
- increase efficiencies and competitiveness;
- reduce time to market and boost productivity.

Damn right they matter!

How much do Brilliant Meetings cost?

Making improvements will not break the bank; every department does not need to have its own boardroom and full-time PA. Much can be achieved through commonsense measures that call upon people to be accountable and act responsibly to themselves and each other; simple measures that are introduced into the organisation's culture with the resolve of senior management to make a difference. For example, how much would it cost to develop a set of guidelines that focus attention on the meeting outcomes, before it even begins, resulting in careful consideration by all involved about the likely meeting effectiveness? *How much will it cost to take NO action?*

quote

'There are risks and costs to a programme of action, but they are far less than the long-range risks and costs of comfortable inaction.'

John F. Kennedy

Without exception all types of meetings should be subject to organisational guidelines, scrutiny and review to ensure productive effective meetings are the norm, not an infrequent, accidental occurrence.

Start today; develop a set of organisational ground rules for all meetings that has everyone's support and willingness to adopt. Right from the outset, participants will understand what the organisation expects from them in the meeting environment.

The Brilliant Meeting effect on employees

Attract, develop and retain key employees

> The ultimate drivers of performance and success within any business are its people

The ultimate drivers of performance and success within any business are its people. When employees are emotionally and intellectually connected, they invest far more of their time and energy into making their employment count, both personally and professionally. Given that so much of knowledge workers' time is spent in meetings, improving meetings will increase the emotional and intellectual connections that employees have with your organisation.

Things that could be implemented now – at little or no cost

- Give responsibility – 'Director of Meetings' – for all meetings to a board member, thereby announcing a new era in meeting effectiveness, and demonstrating that this is being driven from the top.
- Introduce the Brilliant Meetings **ACTION PLAN** (see Chapter 2); ground rules agreed by employees for acceptable behaviour and attitudes before, during and following meetings. Post up a copy of the rules in every meeting room.
- Make Brilliant Meetings a part of every new employee's recruitment and induction process if they are likely to be involved in meetings.

Things worthy of consideration for short-term implementation

- An audit of meeting room resources and technology – projectors, interactive whiteboards, video conferencing, etc., and introduce a training programme that will raise adoption of the equipment that you have already invested in.
- A strategy to ensure that senior executives also fully adopt these resources, so that the drive to be more efficient comes from the top.
- An audit of meeting room effectiveness.
- Training and guidelines to improve the way that PowerPoint™ is used.

Medium-term strategic considerations

Most meetings rooms, irrespective of their size, have a standard configuration of a central table and a number of chairs. Meetings have many varied purposes; they are where people come together to learn, share and create. The room size and structure for collaborating on new ideas is, for example, very different from that required for a formal appraisal or a daily departmental update session, and these should be reflected in your meeting spaces. From the results of your meeting room audit consider the following:

- Use existing meeting spaces better, through configuration changes or enhanced use of existing resources, so that they can deliver better results quickly from effectively supporting differing meeting requirements. Standard meeting rooms could be complimented with collaboration rooms, impromptu rooms and impromptu spaces in corridors and close to coffee machines.
- Introduce a real-time meeting room scheduling system with interactive displays outside each meeting room. This will stop unnecessary interruptions, reduce waste caused by cancelled meetings, and enhance your organisation's reputation.

brilliant tip

For impromptu and short meetings, change the availability of some smaller rooms so that they can be booked for a maximum period of only 30 minutes. This way, these rooms will be available regularly during the day, and you will not have much bigger rooms with expensive facilities being underutilised.

Motivating your employees

Your employees will be motivated by:

- being included in the development of a Brilliant Meeting strategy;
- having a common set of rules by which everyone abides – **ACTION PLAN**;

- having inspiring spaces for their meetings;
- having the correct resources to call upon, for which they have been properly trained and from which they can derive the best value;
- using connecting technologies, such as video and audio conferencing to attend meetings, rather than long road journeys, resulting in a better work/life balance.

---

brilliant example

Collaboration rooms are 'thinking laboratories' where ideas are born, processes developed and problems solved by pooling resources. The resources needed for such environments can be as simple as flip charts, or as technologically advanced as data collaboration solutions but, by matching the resources and process to the meeting type, it will result in an effective outcome for all.

The Brilliant Meeting effect on customers

Increasing customer loyalty and retention

There should be much more to business relationships than just the bottom line price. What organisation today does not profess in its Mission Statement to be 'customer centred'? Yet how many companies take the trouble to consider the interaction with suppliers or clients from the viewpoint of their trading partner?

Things that could be implemented now – at little or no cost

- The Brilliant Meetings **ACTION PLAN.** This will make more effective use of your customers' valuable time and portray an enhanced professional image to them, which can only increase the value of your relationship in their eyes.
- More professional use of PowerPoint™ throughout the organisation.

Things worthy of consideration for short-term implementation

- Guidance for the correct use of (existing) video conferencing to allow more direct and frequent contact with your customers, which enables stronger working relationships to develop.
- The correct use of video conferencing to save everyone's time and costs and ultimately will raise the profile of the organisation as socially responsible.

Medium-term strategic considerations

Introduce 'touch-down' facilities specifically for your top prospects and customers.

What would be the effect on the desire and willingness of that client to work even more closely with you, if there was an area close to reception put aside especially for such valued visitors? An environment where, without disturbing you prior to the scheduled meeting time, they could:

- get a hot or cold drink;
- use the toilet;
- securely hang up their coat and store any bags;
- have wireless access on their own PCs or use a Web-enabled PC and printer whilst waiting;
- sit around the table and go through those last-minute points before the meeting starts.

Of course, given the luxury of space and budget, the concept of a customer 'touch-down' area possibly would feature in every reception area. However, the solution could be as simple as a 'hot desk' with a lockable storage unit.

If such a customer-centred facility was available, it would not matter if your visitor now arrived 30 minutes early. They are not disturbing you, they have somewhere to work or relax and, most importantly of all, their experience of your organisation is just brilliant. It sure beats waiting in a busy, noisy, impersonal reception area, which is a typical pre-meeting experience.

See www.meetingexpert.co.uk for details on customer engagement areas.

The Brilliant Meeting effect on corporate social responsibility targets

More work, less traffic

The 'green agenda' and employee working conditions are now at the top of considerations throughout the planning, manufacture and distribution of goods and services. When it comes to the meeting process, real improvements can be made in all of these areas by choosing to hold meetings over distance; through video, audio and Web conferencing, rather than using carbon-emitting transport and destroying our family routines with excessive, stressful travelling times.

Video conferencing has improved a great deal in the last few years; hardware costs have dropped, secure connection costs using the public Internet are minimal, image and audio quality are now in high definition.

It is not the technology per se that holds back wider usage. Instead it is the meeting environments that contain the equipment, the ease of use brought about through adequate training and the cultural desire to use the technology, that are restricting greater time and money saving. If you already have invested in video, audio and Web-conferencing technologies, then now is the time to take action and make them work for you (see Chapter 13 for more information).

Things that could be implemented now – at little or no cost

- Use the Brilliant Meetings guidelines to determine if a

physical meeting is really necessary every time. If it is not, do not allow it to proceed.

- A process that requires pre-authorisation for approval of travel expenses for attending meetings that could have been successfully completed using video, audio, or Web-conferencing technologies.

Things worthy of consideration for short-term implementation

- Video training to be a mandatory part of employee development programmes, and included as part of the new employee induction process for those employees who are likely to be involved in meetings.

Medium-term strategic considerations

- Introduce dedicated video conferencing rooms that pay full attention to the elements that help drive adoption; necessary to become an intrinsic business tool. Location of the room, audio and lighting quality in the room, and furniture selection and layout. See www.meetingexpert.co.uk.

brilliant example

At the top-end of video conferencing is telepresence; essentially a video conferencing solution that uses rooms exclusively for video meetings. In fact, the room layout is such that you could not use these rooms for non-video meetings. The rooms are constructed to give the best possible audio and visual experience between remote participants, who can all see each other as if they were sitting just three feet across the table.

Whilst these rooms are expensive to buy and run, their value to organisations at board level is now proven, and a wider roll-out within large global organisations is taking place quickly.

Aim to be brilliant – make meetings matter

If we regularly attend boring, ineffective meetings, then that is how we view the organisation that allows this to happen, along with perpetuating negative feelings about our own personal effectiveness.

There is no 'quick fix' available that removes the ingrained, accepted problem of 'bad meetings' but, with a common-sense approach, a change in mind-set and values to accept that meetings are an important function within our job roles, meetings can be made better. Be part of creating a culture that enjoys meetings because of their productivity and effectiveness, and nurture this developing belief that everyone has the responsibility and capability to be part of a Brilliant Meeting every time.

everyone has the responsibility and capability to be part of a Brilliant Meeting every time

brilliant tip

www.meetingexpert.co.uk provides access to many resources associated with making every meeting a Brilliant Meeting. Bookmark this website now, ready for your next Brilliant Meeting: Brilliant Meetings are within our power to create.

Index